Futurist Fantasies

T. K. Van Allen

December 12, 2022

Contents

Introduction

Futurism is mostly science fiction, but it is presented as rational speculation about the future. Futurist ideas are not selected to be realistic or pragmatic. In other words, they are not selected to be predictive of what will actually happen, or prescriptive of what we should do. Instead, they are selected to be emotionally appealing. They are often utopian fantasies, such as economic superabundance, immortality or colonizing the galaxy. Sometimes, they are dystopian nightmares, such as robots taking over the world.

Most futurist fantasies are based on simple extrapolation of recent trends. For example, the global economy and population grew rapidly during the last 200 years. By extrapolating that trend into the future, you might predict economic superabundance and/or a population of trillions expanding into space. Another example is the recent extension of the human lifespan, due to modern medicine, hygiene and nutrition. By extrapolating that trend, you might imagine that people in the future will live for hundreds of years. You might expect us to eliminate cancer and aging, as we have eliminated smallpox and scurvy.

Extrapolating recent trends into the future does not generate good long-term predictions. Trends often exist for a while and then come to an end. An Egyptian in 2600 BC might have predicted that pyramids would just keep getting bigger and bigger. Instead, they went out of style.

Most futurist fantasies are based on the naive assumption that things can keep getting bigger and better forever. The reality is that things can get bigger or better for a while, but progress eventually runs into physical or economic limits. History consists mostly of stagnation. Occasionally, there are brief periods of rapid progress. There are also periods of decline or collapse. Currently, we are in a period of progress, so progress seems natural and inevitable to us, but it won't last forever. In fact, we are probably nearing the end of it.

In this book, I debunk futurist fantasies using basic principles of physics and economics. I also make some predictions of what will happen, and I propose some things that we should do.

This book was originally written as an ebook. As such, it contained many links to external sources. Most were links to Wikipedia pages for important concepts and ideas. Some were links to other sites, such as NASA. For the print version, I have gathered the links together and placed them in a list at the end of the book. I will refer to them as "source 1", "source 2", etc.

Megastructures

The square/cube law is a mathematical law of nature. It has important implications for structures, both natural and man-made. It explains why things don't scale up.

If an object's size is increased on all dimensions by a factor of N, then areas defined within the object (cross-sectional or surface) will increase by a factor of N^2, while the volume of the object will increase by a factor of N^3. That is the square/cube law.

For example, a sphere of radius R has a surface area of $4\pi R^2$, a cross-sectional area of πR^2, and a volume of $(4/3)\pi R^3$. If R is doubled, then the surface area and cross-sectional area both increase by a factor of 4, while the volume increases by a factor of 8.

Generally speaking, strength is proportional to cross-sectional area, while mass is proportional to volume. Thus, the square/cube law implies that mass increases faster than strength. If a structure is scaled up with no change of materials, it becomes proportionately weaker relative to its mass. At some point, it cannot support itself.

The square/cube law explains why structures can't just be scaled up, and it defines practical limits to how big a structure can be.

For example, 12 km is the maximum length of a steel cable that can support its own mass at Earth surface gravity. If it is any longer, it will snap under its own weight. That is true for any thickness, because if you increase the thickness, you increase both the cross-sectional area and the mass by the same factor.

The square/cube law explains why an ant looks different from an elephant. An ant the size of an elephant would collapse and explode. The square/cube law explains why big buildings look different from small buildings. The structure of a skyscraper is very different from the structure of a house. The square/cube law explains why you can kick a toy car across a room without damaging it, while an automobile is badly damaged if it rolls over.

The square/cube law is highly relevant to futurism, because many futurist ideas involve big structures. Imaginary megastructures are common in science fiction and futurism. They are often just scaled-up versions of smaller structures. The square/cube law has specific relevance to space flight, because rockets are big structures that must be very light for their size. The square/cube law is one of the constraints that make space flight very difficult.

Larry Niven's "ring-world" is an imaginary megastructure in science fiction. It is a ring that encircles a star. It revolves around the star faster than the orbital velocity for its radius, fast enough to generate Earth levels of pseudo-gravity on its inner surface. The inner surface of the ring has vastly more land area than a planet, and it absorbs much more of the star's energy.

A Dyson sphere is an even bigger imaginary structure that entirely encloses a star and absorbs all of its energy. A ring-world is just an equatorial slice of a Dyson sphere.

Both structures are physically impossible, because of the square/cube law. No known material could sustain the forces generated by the ring's rotation. Just because we can make a hula hoop, it doesn't follow that we can make a ring the size of a planetary orbit. I read Ringworld when I was young, and I

4

enjoyed it, but I was blissfully unaware of the square/cube law. It isn't really possible to have structures that big.

Above a certain size, structure ceases to exist, and the shape of an object is determined only by gravitation and rotation. For example, a cube the size of the Earth could not exist. It would collapse into a sphere, due to gravitational forces. A large ring can exist, but only as a collection of smaller objects in orbit together. For example, Saturn's rings consist of dust and rubble orbiting Saturn.

Now, let's consider a structure that is more down to Earth but still pie in the sky: the space elevator or space ladder. It is an important concept in futurism, as a way to make space travel more energy efficient.

The idea is to have a very long cable that is anchored to a point on the Earth's surface and to an object in space. The center of mass of the system as a whole (cable and space-anchor) must be slightly beyond geostationary orbit, so that the centrifugal force of rotation slightly exceeds the force of gravity. That would keep the cable taut, while maintaining its up-down orientation relative to the anchor point on the Earth. Geostationary orbit is roughly 36,000 km from the Earth's surface, so a very long cable would be necessary. The cable would have to support the tension created by gravity and centrifugal force acting on the mass of the cable and its payload.

Conventionally strong materials, such as steel, are much too weak for such a cable. A steel cable longer than 12 km would break under its own weight. We could increase the width of the cable as it goes up, so that the cable above could support the mass of the cable below. Or we could double the number of cables periodically. However, this would create an impossibly large structure.

For steel, the cable structure would have to be 1.6×10^{33} times thicker at geostationary orbit than at the Earth's surface. (See source 4.) For comparison, the Sun's diameter is roughly 1.4×10^9 m, the diameter of the Milky Way galaxy is roughly 1×10^{21} m, and the diameter of the observable universe is roughly 1×10^{27} m. A steel cable that is 1 mm thick at the Earth's surface would have to be much wider than the observable universe at geostationary orbit. This is obviously impossible.

A space elevator or space ladder would require materials that are much stronger than steel. There are some hypothetical materials that are strong enough, such as carbon nanotubes, but they are difficult and expensive to make.

There are many other problems with space elevators, which I will discuss in the next two chapters. Unfortunately, space elevators are likely to remain science fiction forever.

Now, let's consider a favorite idea of both futurism and science fiction: the rotating space habitat. Unlike a ring-world or a space-elevator, this megastructure is physically possible using conventional materials. But would it be economically feasible?

The O'Neill cylinder is one design for a rotating space habitat. It was proposed by a physics professor and his students back in the 1970s. I couldn't find a detailed version of the original design, so I will present a simplified version. My simple design is a cylinder 10 km long and 6.4 km in diameter, with a 1-m-thick hull. It would be rotating at a speed that would create 1 g of pseudo-gravity on the inner surface.

This is physically possible, given the strength of steel. According to my calculations, the maximum size of a steel cylinder with 1 g of pseudo-gravity is 24 km in diameter. The cylinder would be held together by tension in the hull. For holding itself together, the hull's thickness doesn't matter, because the strength and the mass are both proportional to the thickness (assuming that the thickness is much smaller than the radius of the cylinder). However, the hull needs to do more than just hold itself together. It must support the content of the cylinder, which adds mass but not strength. It should provide protection from radiation. And we want the hull strength to have a comfortable margin of error. So, a hull that is 1 m thick and 6,400 m in diameter seems reasonable.

Suppose that we are going to build this structure in low Earth orbit (LEO).

How much steel do we need? The surface area of a cylinder is given by the formula:

$$A = 2\pi RL + 2\pi R^2$$

For our cylinder, R = 3,200 m and L = 10,000 m. Plugging those numbers into the equation, we get roughly 265 million square meters. For a 1-meter-thick hull, we would need 265 million cubic meters of steel. The density of steel is roughly 7,700 kg/m^3. So, we would need about 2 trillion kg of steel.

Of course, there's no point having an empty cylinder in space. We would want to put something in it, such as air, water, buildings, etc. Let's suppose that we want to put 1 trillion kg of material inside the cylinder. In total, we would need to lift 3 trillion kg to LEO.

Currently, the cost of putting 1 kg into LEO varies between $50,000 and $1,500. SpaceX claims the lowest cost, for the Falcon Heavy. (See source 6.) At the time of writing, the

Falcon Heavy has only completed 3 launches, so that claim could be somewhat optimistic. For this thought experiment, we'll use a hypothetical cost of $1,000/kg. That is lower than SpaceX's optimistic estimate for the Falcon Heavy, and much cheaper than other existing vehicles.

At $1000/kg, the cost of putting 3 trillion kg into LEO would be $3 quadrillion. The current global GDP is about $100 trillion. So, it would cost 30 times the current global GDP just to bring the materials to the construction site.

What would be the return on that investment?

The structure would have roughly 200 km^2 of inner surface area. Perhaps we could create a mini-biosphere in this steel can in space, and have some people living inside it, farming crops, etc. Currently, the average cost of farmland in the US is $3,160 per acre, which is about $780,000 per km^2. That is for farmland in one of the most agriculturally productive countries in the world, where the sun shines and water falls from the sky. Even if we round up to $1 million per km^2, the value of the land in our hypothetical O'Neill cylinder would be only $200 million.

Of course, that's an absurdly optimistic estimate. Farmland in space would have no value.

Here are some things that would be much easier than colonizing LEO:

- Irrigating the Sahara, or any other desert area, to grow crops.

- Growing food in greenhouses at high latitudes.

- Building floating islands in the ocean.

- Colonizing Antarctica.

- Colonizing the deep-sea floor.

- Building cities deep underground.

We aren't doing those things now, because they would just waste resources. Building a space habitat would be a much bigger waste of resources. No problem that exists on the Earth would be solved by moving a few dozen people into a steel can in space.

The O'Neill cylinder is a fun thought experiment, but it is economically insane.

What about a smaller structure, such as a wheel 500 m in diameter and 10 m across? That would be much cheaper to build, of course. But it would still be very expensive. Unlike the bigger cylinder, it couldn't contain a self-sustaining ecosystem. It would always require resupply from the Earth. Would it be economically viable as a hotel or a research station? I don't think so. The International Space Station (ISS) cost roughly $100 billion to build, and it costs $3 billion per year to operate. (See sources 7 and 8.) It provides us with little more than pictures of the Earth.

Big structures are hard to build, and they have diminishing returns — especially if we build them in space.

Space Flight: Ascent

Going to space is very hard, because of the rocket equation and the square/cube law.

The Tsiolkovsky rocket equation gives the ratio of initial mass to final mass as a function of ΔV, the change of velocity. Actually, it's a bit more complicated than that. The equation also depends on exhaust velocity. Here it is:

$$M_0 / M_F = \exp(\Delta V / V_E)$$

Equivalently:

$$\Delta V / V_E = \ln(M_0 / M_F)$$

M_0 is the initial mass of the rocket. M_F is the final mass, after the propellant has been expelled. ΔV is the change in velocity. V_E is the exhaust velocity: the velocity of the propellant leaving the rocket engine, relative to the rocket.

On the Earth, we normally move around by pushing against matter in the local environment. Feet push against the ground. Wheels push against the road. Propellers push against air or water. In space, there is nothing to push against. Rockets propel themselves by throwing matter out the back.

For simplicity, we can think of ΔV as a change in speed. However, a rocket does more than just accelerate. It also does work against the opposing forces of gravity and air friction. To include those effects, ΔV is defined as the integral of thrust divided by mass with respect to time. ΔV quantifies the net effect of a rocket engine.

For a fixed V_E, the ratio M_0 / M_F grows exponentially as a function of ΔV. Thus, the ratio of propellant mass to total rocket mass approaches 100% very quickly. This is called "the tyranny of the rocket equation".

The essence of the rocket equation can be understood without math. Propellant is used to lift and accelerate a payload. But the propellant must also be lifted and accelerated. So, we need more propellant to lift and accelerate the propellant, and then we need even more propellant to lift and accelerate that propellant, and so on. Most of the propellant in a rocket is used to lift and accelerate the propellant. That's why we need a huge rocket to put a satellite into orbit or send a space probe to Mars. The payload is much smaller than the rocket.

For example, the mass of the Saturn V rocket was 85% propellant, 10% structure, and 5% payload. (See source 11.) We can contrast that with other modes of transport. A car is only about 4% fuel by mass. A fighter jet is about 30% fuel. Cars and fighter jets also have the advantage that they can get oxygen from the atmosphere.

Rockets have to be big, and most of their mass has to be propellant. That's where the tyranny of the rocket equation meets the tyranny of the square/cube law. Rocket design is constrained on one side by the rocket equation, and on the other side by the square/cube law.

Due to the rocket equation, rockets must be big and mostly propellant. Due to the square/cube law, as we increase the size of a structure, either the structure becomes weaker, or we need to dedicate more of the volume and mass to the structure itself, and less to the content. Rockets need to be big and strong, while dedicating only a small percentage of their total mass to the structure. That makes rocket design a very hard problem.

Rocket designers use special materials and construction techniques to make the structure both light and strong. The Space Shuttle external tank is a good example. Although it looks boring (just a big cylinder with rounded ends), it was a marvel of design. It had a very high ratio of content to total volume: about 96%. That is comparable to a soda can, which is another marvel of design. However, because of the square/cube law, it was much harder to attain that ratio for the external tank, and the tank was much more fragile than a soda can. You can drop a full soda can on the floor, and it won't break or explode. The same could not be said of the Space Shuttle external tank.

Due to design constraints, rockets are fragile. They operate close to physical limits.

The Space Shuttle Challenger Disaster

On January 28 1986, the Space Shuttle Challenger exploded shortly after lift-off, killing everyone on board.

The disaster was caused by the failure of a seal on one of the solid rocket boosters. Exhaust gases leaked through the seal and damaged the external tank, causing a rupture. After the tank lost its integrity, it exploded.

There were concerns about that seal before the launch. It was known to be weak. Unusually cold temperatures for Florida might have contributed to the disaster, by reducing the flexibility of the materials used in the seal.

The Challenger disaster is a good example of how rockets are *fragile*. They are very close to the limits of what is physically possible, because their design is highly constrained. Normally, engineers and architects design structures that are *robust*: they can withstand significant deviations from normal conditions. With rocket design, this is not possible.

There were also social reasons for the disaster. Engineers were concerned about the seal and the effects of cold temperatures on it. They advised delaying the launch, but their concerns were overridden by management. Managers tend to be optimistic, because they need to promote their organization and what it does. Optimism can have tragic consequences.

It was a high-profile mission, which might have affected the decision to launch. The crew included the first civilian astronaut, Crista McAuliffe. She was part of the "Teacher in Space" project, which was intended to inspire students to study science and technology. There were many public interest stories about her and her class before the launch. To a large extent, the mission was a publicity stunt.

Over the years, NASA has portrayed space flight and exploration in mystical, quasi-religious terms: as a necessary and inevitable aspect of human progress. Going to the Moon was "one giant leap for mankind", although it's not clear toward what. The Space Shuttle program was motivated by this quasi-religious view of space travel, not by pragmatism.

To be fair, NASA is in a difficult position. Public expectations of space travel have been shaped by science fiction. In Star Trek, every problem is solved by a combination of magical technology and the human spirit. The reality of space travel is very different. It is difficult and dangerous, and it doesn't always have a happy ending.

Rocket Engines and Propellants

There are various types of rocket engines and propellants. Most rockets use liquid fuels, such as liquid hydrogen or kerosene, combined with liquid oxygen. Liquid hydrogen and liquid oxygen must either be stored under high pressure or kept very cold (cryogenic). High pressure tanks need to be very thick and strong, and thus heavy, which is not feasible for a rocket. So, most rockets use cryogenic propellants.

Typically, cryogenic propellants are pumped into the rocket just before launch, and rapidly consumed during the ascent. They cannot be stored for very long, because they warm up and boil off. Propellant tanks have vents so that boil-off can escape. Without venting, pressure would build up in the tank.

Propellant loss is a minor issue for short-term storage, but it becomes a major issue for long-term storage. Another issue is that different propellants require different temperatures, which makes storing them in close proximity more difficult. Hydrogen requires the lowest temperature, about 20°K.

Cryogenic propellants are good for ascent, but they are problematic for long-distance travel, due to storage issues.

Solid propellants have their own advantages and disadvantages. Unlike cryogenic propellants, they can be stored ready-to-use at normal temperatures. That is especially important for military applications, such as missiles. Solid propellants have two disadvantages: they cannot be controlled after ignition, and they are less powerful than liquid propellants. Thus, they have limited uses in space flight. Notably, the Space Shuttle used solid rocket boosters for the first phase of its ascent.

There are also hypergolic propellants, which are liquid at normal temperatures and do not require ignition. They consist of two liquids that ignite when mixed. Hypergolic propellants are useful for long-distance space travel, because they can be stored for a long period of time, and the engine can be easily started and stopped. Their main disadvantage is that they are less energy-dense than cryogenic propellants. The Apollo service module, which carried astronauts to the Moon, was powered by hypergolic propellants.

A typical liquid-propellant rocket engine burns kerosene, hydrogen or methane with liquid oxygen. It pumps those liquids into the engine rapidly, to generate a high thrust. Cryogenic propellant is also used to cool the engine, to prevent melting or burning. The temperature inside a rocket engine is too high for most materials. For solid rockets, the nozzle ablates (burns/melts away) during use.

There are many challenges involved in rocket engine design. One is how to pump propellants into the engine fast enough. Another is maintaining a steady flow of propellants. During ascent, inertia and gravity help to pull the propellants toward the bottom of the tanks. In free-fall (such as orbit), pumps must do all the work. Sloshing can be a problem if the propellant tanks are partially empty.

Alternatives to Rockets

Are there any viable alternatives to rockets for space flight? The simple answer is no. There are some proposed alternatives that don't violate the known laws of physics, such as space elevators and mass drivers, but they have major problems.

In Megastructures, I described the problems with space elevators that are due to the square/cube law. There are other problems.

For example, a space elevator could deliver a payload to geostationary orbit or beyond, but not to low or medium Earth orbit. If you got off a space elevator at LEO, you would just fall back to the ground. To put an object into LEO, the object must be accelerated to an orbital velocity, which is much faster than the speed of the Earth's rotation.

Another problem is that the elevator would cross through the Van Allen belts, which contain high levels of particle radiation. A rocket can take a trajectory that reduces radiation exposure (at higher latitudes), and rockets travel very fast. A space elevator would have to traverse the thickest part of the Van Allen belts, while moving at a much slower speed.

There are many other problems with the space elevator concept, such as wind, collisions with satellites, the enormous expense of building one, etc.

Futurism has no shortage of crazy ideas. An orbital ring is an imaginary megastructure that would support space elevators at LEO. It would be a giant ring encircling the Earth, rotating at a velocity that is slightly above the orbital velocity for its altitude. The rotation would generate enough centrifugal force to support elevator cables. An orbital ring has the usual problems with megastructures, and it would be insanely expensive.

What about a mass driver? A mass driver is a hypothetical device that accelerates a payload using a linear motor. Essentially, it is a long tube, like a gun. The payload accelerates to a high velocity inside the tube, and then flies into space. That's the idea, anyway.

There are many problems with the mass driver concept. If the tube contained air, air friction would be a problem. It would slow down the payload and make it very hot. Ideally, the payload would be accelerated in a vacuum, but that would only work if the tube's exit was in space. The payload would be destroyed if it entered the Earth's atmosphere at an orbital velocity from a vacuum. If the exit was in space, the gun would need to be at least 50 km high. It would be impossible to build with conventional materials, due to the square/cube law. There is also the problem of maintaining a vacuum in a giant tube.

A mass driver could be useful in space or on the moon, but it would not be feasible on the Earth.

For now, there are no better alternatives than chemical rockets, and chemical rockets aren't very good.

There is a lot of hype about space flight these days, mostly due to quasi-private companies marketing their products. (SpaceX comes to mind.) However, there hasn't been much progress in space flight since the 1970s. The lack of progress is due to hard physical constraints: the rocket equation and the square/cube law. There isn't much room for progress.

In science fiction, a spaceship (such as the Millennium Falcon) can gently lift off from a planet, fly into space, and then gently come back down. That would require a radically new form of propulsion, such as anti-gravity, which would require some new laws of physics. So, it's likely to remain a fantasy.

Space Flight: Descent

Going to space is difficult. Coming back from space is even more difficult — if you want to get back in one piece.

To put an object into deep space or LEO, you need to put a lot of energy into it. An object in orbit has a huge amount of energy relative to the same object at rest on the Earth's surface. Suppose that the object is orbiting at an altitude of 200 km (a typical LEO). At that altitude, the orbital velocity is 7,790 m/s. The energy of the object has two parts: kinetic and gravitational.

- Kinetic: $M \times V^2 / 2$

- Gravitational: $M \times A \times H$

In these formulas, M is the mass, V is the velocity, H is the height, and A is the gravitational force per kg (equivalently, the acceleration rate due to gravity). The second formula is a simplification, because the force of gravity is a function of height. However, it doesn't change that much between the Earth's surface and 200 km up. So, I'll use the simple formula with $A = 9.6$ Newtons/kg.

Given those numbers, the energy per kg is:

- Kinetic: 30.34 megajoules

- Gravitational: 1.92 megajoules

- Total: 32.26 megajoules

This is how Wikipedia describes a megajoule:

The megajoule (MJ) is equal to one million (10^6) joules, or approximately the kinetic energy of a one megagram (tonne) vehicle moving at 161 km/h (100 mph).

For our hypothetical object in LEO, each kg of mass has roughly 32 megajoules of energy, of which 30 are kinetic and 2 are gravitational. That energy must be dumped before the object can be at rest on the Earth's surface.

That's why it is difficult to return from space in one piece. The returning object must safely dump a huge amount of energy.

Currently, aero-braking is the only way to do that. The energy is dumped into the atmosphere as heat created by air friction. The main problem with aero-braking is that it generates very high temperatures. For a vehicle to survive re-entry, it must be able to withstand those temperatures without losing its structural integrity. The drag forces can also be dangerous. Aero-braking requires a strong, heat-resistant structure.

So far, two approaches to re-entry have been tried: the space capsule and the space plane.

A space capsule is a container designed to survive re-entry. It is shaped like a compressed raindrop, and it falls with the wider bottom surface facing down. The bottom is designed to gradually ablate (melt/burn away). The ablated material carries away heat. The capsule is insulated to protect passengers and cargo from the heat, and it is strong enough to withstand drag forces.

After the capsule has fallen to a relatively low altitude and slowed to a relatively low speed, parachutes are deployed, causing it to decelerate further. It then splashes down in the ocean, where it is retrieved by a boat. (Some of the early Soviet capsules landed on land, rather than splashing down in the ocean.)

A space capsule must be relatively small, due to another version of the square/cube law. A smaller object decelerates faster as it falls, because the force of air friction is proportional to cross-sectional area, while the force of gravity is proportional to mass and thus to volume. As a structure is scaled up, its mass increases faster than its cross-sectional area.

A space capsule is much smaller than the rocket that puts it into space. It can carry only a few passengers and/or a small amount of cargo back to the Earth. A large capsule would fall too fast and create too much heat. It would partially burn up and then crash.

The other approach is the space plane, of which the Space Shuttle is the best example. The Space Shuttle was intended to be the next step in space flight: routine travel to and from space. Previously, manned space flight involved sending a huge rocket to space, and having a small capsule return. Nothing was reused. In science fiction, a spacecraft (such as the Millennium Falcon) can simply fly into space and fly back down, like an airplane crossing the Atlantic. The Space Shuttle was an attempt to make space flight more like science fiction, with a vehicle that could fly into space (with the assistance of a massive external tank and two huge rocket boosters) and then fly back down.

The Space Shuttle used a special technique to return to the Earth. It would first reduce its velocity by firing its engines in the direction of travel, causing it to fall out of orbit. It would then orient itself at a 40° angle, to increase the air friction on its lower surface. The wings increased the surface area and created lift, which slowed the descent. The shuttle also executed S-curves in the upper atmosphere, like a skier zig-zagging downhill, to slow the descent. The gradual descent allowed it to dump energy over a longer period of time without getting too hot. Nevertheless, the surface still reached very high temperatures, up to 1500°C. The shuttle did not use ablation to carry away heat. Instead, it was covered in special ceramic tiles that could withstand extremely high temperatures. When it reached a lower altitude and speed, it would fly like a glider to the landing site, where it would land on a runway and deploy parachutes to slow down to zero.

The Space Shuttle was not quite the Millennium Falcon, but it was a valiant effort in that direction.

The Space Shuttle program lasted from 1981 to 2011. There were 135 missions, of which 133 returned safely, and 2 were disasters. I have already described the first disaster, which occurred when the shuttle was going up. The second disaster occurred when the shuttle was coming down.

The Space Shuttle Columbia Disaster

In 2003, the Space Shuttle Columbia disintegrated during re-entry, killing everyone on board.

The disaster was due to damage that occurred during lift-off. Some insulation fell from the external tank and hit one of the shuttle's wings, damaging it. The impact was observed from the ground, but the damage was assumed to be minor, and re-entry went ahead as planned. During the descent, hot air leaked into the damaged wing, destroying its structural integrity. The wing tore apart, and then the uncontrolled drag forces caused the shuttle to disintegrate.

Again, space flight involves fragile structures operating very close to their physical limits. A small accident can cause a catastrophic failure.

Alternatives to Aero-Braking

There are no feasible alternatives to aero-braking. If your expectations have been created by science fiction or SpaceX marketing videos, you might believe that propulsive braking is an option, but it really isn't. It would require a huge amount of propellant to slow the descent of an object from space, and the propellant would need to be carried up during the ascent. Given the huge costs of sending anything to space, it is not pragmatic to send up propellant instead of payload.

SpaceX uses propulsive braking to land the first stage of its rockets, but the first stage doesn't reach an orbital altitude or velocity. When it begins its descent, its velocity is zero. It stops going up, and then it starts falling down. It only has to lose a relatively small amount of energy, so it doesn't reach a high temperature. Landing a first stage is nothing like returning from orbit.

Reuse is an economic trade-off. On the one hand, it is cheaper to refurbish an existing rocket than to build a new one. On the other hand, recovery requires additional mass (propellant or a parachute), which reduces payload mass and/or ΔV. Recovery has other costs, such as retrieving and transporting the recovered vehicle. There could also be an increased risk of failure for reused parts.

It is worth mentioning SpaceX's badly named "Starship", which is analogous to the Space Shuttle. Starship is designed to go to space and return, using a combination of aero-braking and propulsive braking to land. At launch, it would sit on top of a first stage (the "Super Heavy" booster), which is also recoverable. Like the Space Shuttle, Starship has a coating of special ceramic tiles to withstand the temperatures of re-entry. It also has wings, of a sort, which are used to steer it and slow its descent. Unlike the Space Shuttle, however, Starship has a very limited ability to maneuver. It cannot fly or glide. It just falls, although it can be steered to some extent. So, it will not use S-curves to slow down, and it cannot glide to its destination and land on a runway. At the time of writing, it has yet to complete a successful test flight. It remains to be seen if it will actually work.

What about a space elevator/ladder? Could we just climb back down from space? The difficulty of creating a space elevator/ladder was addressed in <u>Megastructures</u>. But let's imagine that we have one. Before you could climb down it, you would need to be at rest relative to it. Thus, you would need to match its orbital velocity. If you were returning from LEO or deep space, that would be a huge ΔV. A space elevator or ladder would only be useful for a spacecraft in geostationary orbit.

Using a mass driver to descend from space is even less feasible. It would be like trying to steer a bullet into the barrel of a gun on a perfect trajectory. We can imagine ways of doing that (an electromagnetic funnel?), but it's not likely to ever exist.

Unless we discover anti-gravity, the Millennium Falcon will remain a fantasy. We'll probably be stuck with chemical rockets and aero-braking as the only feasible methods of ascent and descent. Since we are already operating near the physical limits of those technologies, I don't expect much progress in space flight. We might have already reached the limits of what we can do.

Was the Space Shuttle "Peak Space"?

For someone who likes the idea of space flight and space travel, it is depressing to consider that our space flight abilities might have already peaked.

The Apollo program first went to the Moon in 1969. The last manned lunar mission was in 1972. We stopped going to the Moon for two reasons:

- It is extremely expensive and dangerous.

- It has no economic benefits.

When something has huge costs and no benefits, we're not likely to make a habit of doing it. The only reason to go to the Moon was to show that we could, and we've already done that. There is no reason to go back.

After the lunar missions, NASA's focus shifted to unmanned exploration of the solar system, and to the Space Shuttle, with the intent of making manned space flight more like air travel: safe and routine. Unfortunately, that goal was not feasible.

The Space Shuttle had capabilities that no existing vehicle has. It could carry both passengers and cargo to and from space on a regular basis. The space-plane was reusable, although it needed considerable refurbishing between flights. In many ways, it was a triumph of engineering. However, there was no need for such a vehicle. To put a satellite into orbit, or send a space-probe to Jupiter, it is more efficient to use an unmanned multi-stage rocket. Manned space flight is romantic, but not practical.

In some ways, the Space Shuttle program was very successful. It launched 135 missions, with only 2 failures. It routinely put men into space. For a while, it captured the public's imagination. But it filled a need that did not exist.

The Space Shuttle program was based on the expectation that we would explore and colonize space. Space was the "final frontier". Exploring space was the next step in human progress. This view was neither realistic nor pragmatic. It was quasi-religious.

Space exploration was in the zeitgeist of the 1970s and 1980s, partly due to the space race, and partly due to popular science fiction, such as Star Wars and Star Trek. However, the reality of space flight didn't live up to the expectations created by science fiction. The public eventually got bored of the Space Shuttle program, and it was quietly retired.

Asteroid Mining

In this chapter, I will explain why asteroid mining is a crazy idea, and we are never going to do it. I will also discuss the various challenges involved in long-distance, manned space travel, such as going from the Earth to an asteroid and back. I will use a hypothetical asteroid-mining mission as an example.

Suppose that we (two miners) are planning to travel to an asteroid located between the orbits of Mars and Jupiter. There, we intend to mine for precious metals, which we will bring back to the Earth. The round trip will take roughly 1,000 days.

Let's consider the many problems with such a venture.

First, it has all the problems of ascent and descent. I described those problems in previous chapters, so I won't address them in detail here. But they must be solved, and they add considerable risk and expense.

Most problems of space travel depend on mass. The bigger the payload, the bigger the problems. It requires a huge amount of energy to put a small payload into space and accelerate it to Earth escape velocity. A rocket's mass is mostly propellant, due to the rocket equation. A small increase in the payload size requires a large increase in the rocket size. But rockets are hard to scale up, due to the square/cube law. Thus, conveying a large payload to an asteroid and back is a very hard problem.

We need a large payload. We need supplies and equipment to sustain ourselves. We need propellant for the trip there and back. We need comfortable living quarters and some amenities, because it will be a long journey. We need to protect ourselves from radiation. We are going to bring back cargo.

Now, let's consider some of the specific problems involved in long-distance space travel:

- **Life support:** We need oxygen, water, food, waste disposal and electrical power. Our life-support supplies and equipment will add a lot of mass.

- **Radiation:** Space is full of dangerous radiation. We need shielding to protect us, which will add more mass.

- **Micro-gravity:** The human body is adapted to gravity. Low or no gravity is harmful. Generating pseudo-gravity by rotation has many issues.

- **Thrust:** We need an engine and propellant to accelerate, decelerate and maneuver in space. Again, this will add mass to the vehicle. And the rocket equation applies: we need propellant to accelerate propellant.

I'll describe each problem in more detail.

Life Support

According to NASA, an astronaut consumes roughly 6 kg of material per day, which includes (roughly) 1 kg of oxygen, 3 kg of drinking water, and 2 kg of food (which includes some water). (See source 23.) Let's assume an additional 1 kg per day of packaging and miscellaneous material. With no recycling, an astronaut would need 7,000 kg of supplies for a 1,000-day mission.

Let's compare that to the payload capacity of the Space Shuttle. It could carry 27,500 kg to LEO (200 km altitude), 16,000 kg to the ISS (400 km), and 2,300 kg to geostationary orbit (35,800 km). Again, it takes a huge amount of propellant to deliver a small payload to space. Our supplies alone would be roughly the same mass that the Shuttle could deliver to the ISS, but we are going far beyond LEO. We need to accelerate that payload to Earth escape velocity, and then accelerate even more to get to a higher solar orbit (beyond Mars). And we need to bring more than just supplies.

Can we use recycling to reduce the amount of supplies? Perhaps, but recycling is not a free lunch. It requires equipment and energy. It also adds risk. If the recycling equipment failed, we would die without oxygen, water or food.

We need an electric generator of some kind. Space probes use either solar panels or radioisotope thermoelectric generators (RTGs). Solar panels don't generate much electricity, and they add significant mass. The same is true for an RTG, and it would also add significant risk. An RTG is powered by a highly radioactive substance, such as plutonium. There is obviously a risk in sending an RTG into space. An accident during ascent or descent could contaminate the Earth. There have been some accidents in the past, and NASA now requires RTGs to be capable of surviving re-entry intact. Another possibility is a compact nuclear reactor, but it would be more massive than an RTG and add even more risk.

After perusing NASA documents for a while, I didn't find a good solution to the electricity problem. They recognize the need for a reliable, high-output electrical generator, but they don't propose any specific solution.

The Space Shuttle and Apollo programs used hydrogen fuel cells to generate both electricity and water. That approach works for short missions, but it requires storing hydrogen and oxygen, which are difficult to store in large quantities for a long period of time. They would need to be in liquid form, which requires either high-pressure tanks or very cold temperatures. Also, hydrogen tends to leak out of any container.

We can imagine other approaches, but they all have problems. Batteries are not very energy-dense, and they lose energy over time. An oil-burning generator would require storing a large amount of fuel oil and oxygen. Storing the oil would not be a huge issue, but storing the oxygen would be a problem.

Radiation and Collisions

On the Earth, we are protected from radiation by the atmosphere and the magnetosphere. In LEO, astronauts are still protected by the magnetosphere, but they are exposed to much higher levels of radiation than on the Earth. There are also high-radiation zones within the magnetosphere: the Van Allen belts. Astronauts passing through those zones are exposed to higher levels of radiation for a short time. In deep space, beyond the Earth's magnetosphere, the background radiation is much higher than it is on the Earth.

The following chart shows radiation exposure in milli-Sieverts for various events. On average, a US citizen receives roughly 6 mSv per year from natural and artificial sources. During 6 months on the ISS, an astronaut receives about 75 mSv. During a 180-day trip to Mars, an astronaut would receive about 300 mSv. During 500 days on Mars, where there is some shielding by the atmosphere and the planet itself, an astronaut would also receive about 300 mSv. (See sources 25 and 26.)

Those levels are not immediately deadly, but radiation is harmful in the long run. It damages DNA and other micro-structures within cells, causing cell death, dysfunction and cancer.

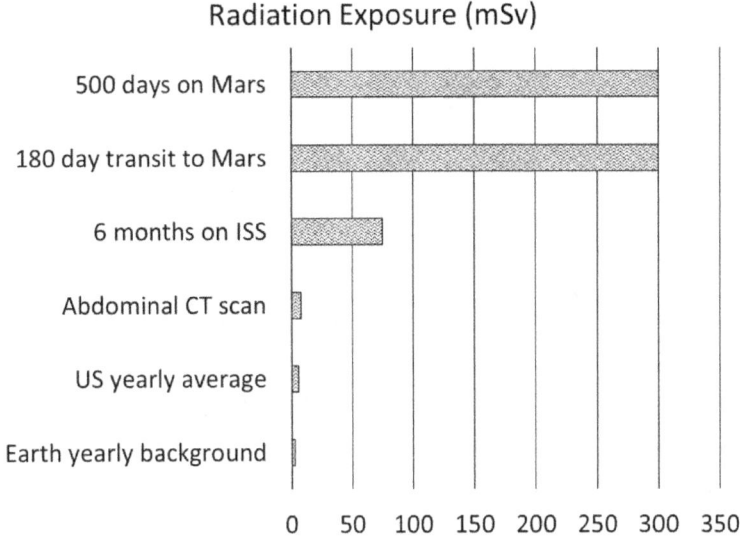

Radiation Exposure (mSv)

Shielding can reduce radiation exposure, but it requires adding more mass to the vehicle. Some of the radiation in space consists of high-energy particles called "cosmic rays". They were once thought to be photons, but are now known to be particles, such as protons, helium nuclei and occasionally the nuclei of heavier elements, such as carbon or iron. Cosmic rays require very thick shielding to block, and they create secondary radiation when they collide with large atoms, such as iron or aluminum. Liquid hydrogen and water make good shielding material for cosmic rays, and they have other uses, but they would have to be stored in the hull (or in some way that makes a good shield), and they would have to be replaced as they were consumed.

There is also the risk of collisions with meteoroids. Even colliding with a grain of sand is dangerous at the speeds involved in space travel. That danger could be increased by mining activities, such as blasting.

31

Micro-Gravity

The human body is adapted to Earth gravity. A number of health problems are caused by extended stays in low gravity environments.

Normally, the heart works against the force of gravity to pump blood. Without this opposing force, blood can pool in the upper parts of the body, causing swelling. The brain, eyes and ears can be affected. In space, you need less blood. The body compensates by reducing blood volume, which is a problem when you return to the Earth. The skeletal muscles also atrophy, because they do less work. Bones lose density, and the calcium that is pulled out of bones creates a risk of kidney stones.

Pseudo-gravity can be generated by rotating the spaceship, or by having a rotating structure inside it. But there are some issues with that.

First, a large radius of rotation is necessary to generate Earth-like gravity without inducing vertigo. Rotation rates above 1 rpm cause problems. To have Earth-like gravity at a rotation rate of 1 rpm, the radius of rotation must be almost 1,000 m. A large rotating structure requires a lot of mass, for obvious reasons. Even a small rotating structure would require significant mass.

Rather than rotating part of the ship, we could rotate the entire ship. However, that would create other problems. It requires energy to start or stop the rotation. The angular momentum would make the ship less maneuverable. (You can experience angular momentum by holding a spinning bicycle tire in your hands and then tilting it.)

Pseudo-gravity is not feasible, unless we have a very big spaceship and plenty of energy to spare.

Thrust

Our spaceship needs engines to accelerate and decelerate. It will have to push against the Sun's gravity to reach a higher solar orbit. We might want to use thrust to accelerate to a higher speed, so that we reach our destination faster. In that case, we will also need to decelerate when we get there. So, our spaceship will need some type of propulsion, which requires some type of propellant. Of course, that adds more mass to an already massive vehicle. It also adds some design challenges.

What type of engine and propellant should we use?

Cryogenic propellants (such as liquid oxygen, hydrogen and methane) are problematic for long-distance travel, because they must be stored at very low temperatures.

A spaceship in the solar system, lit by the Sun, will stay relatively warm, although it will be warmer on the sunlit side and cooler on the dark side. Of course, its temperature will depend on its distance from the Sun.

Space doesn't really have a temperature, in the way that air has a temperature. In space, the temperature of an object is determined by the electromagnetic radiation that it emits and absorbs. At a certain temperature, there will be a balance between emission and absorption. At the Earth's distance from the Sun, the "effective temperature" is about 6°C for an object that absorbs all solar radiation. (See source 28.) That is a rough approximation to the equilibrium temperature for an object near the Earth. If the object is completely shielded from sunlight and other local radiation sources, the equilibrium temperature is about −270°C or 3°K. That is the "temperature" of deep space.

It is possible to keep things very cold in space with appropriate shielding and insulation, but it would not be easy for a large fuel tank. The tank would absorb heat from the rest of the ship via conduction and infrared radiation. Of course, shielding and insulation would add more mass to the ship.

When cryogenic propellants are used during ascent, they are stored for a brief period of time, so warming isn't an issue. Neither is leakage. The storage tanks are often vented, so that boil-off can escape. On a long-term voyage, however, the propellant would warm up, and venting would cause a significant loss of propellant, perhaps a complete loss. I tried to find some information about the long-term storage of cryogenic propellants, but I couldn't find much.

The Apollo missions to the Moon used hypergolic propellants, which are liquid at normal Earth temperatures. They react when mixed, so there is no need for an igniter. Two liquids are sprayed into the combustion chamber, where they spontaneously ignite and generate thrust. There are some issues with storing hypergolic propellants for long periods of time. They are highly toxic and corrosive. Also, they are less energy-dense than standard cryogenic propellants.

Solid propellants are safe and easy to store, but they cannot be controlled after ignition. Thus, they cannot be used for precise or unplanned maneuvers. They are also less energy-dense than standard cryogenic propellants.

Nuclear engines have been proposed for long-distance space travel, but they have several disadvantages. A nuclear engine would be very big. It could contaminate the Earth if there was an accident during ascent or descent. It would need radiation shielding and a cooling system, adding yet more mass. It would be dangerous if it broke down, and it would be difficult to fix. It would require some type of propellant to heat and spray out. Hydrogen is the most energetically efficient propellant, but it is difficult to store. Perhaps a less efficient propellant, such as water, could be used. A nuclear engine could solve the thrust and energy problems of long-distance travel, but it would create many new problems.

Mining the Asteroid

Suppose that we have solved the problems of space travel, and we have arrived at our destination after 450 days in space. We now face a new set of problems.

First, how do we land on the asteroid, or anchor the spaceship to it?

On the Earth, we have friction to slow us down, and gravity to anchor us to the ground. It might be very difficult to land on an asteroid with low gravity and no atmosphere. We would need to carefully match the velocity of the asteroid in space, so that we are almost stationary relative to it. Then we would slowly descend to the surface. We would probably need to use some thrust to control the descent. The engines might create a cloud of dust, which would not settle quickly, as it does on the Earth. If the surface is soft, the ship might sink into it. If the surface is hard, the ship might bounce off it. We can't assume a perfectly flat landing site. We would need to be very careful not to damage the ship, which is our transportation back home. Also, the spaceship would need "legs" to cushion the landing and support it on the surface, which adds yet more mass to the vehicle.

Suppose that we manage to safely land the spaceship on the asteroid. Now, the mining begins! Hi ho, hi ho, it's off to work we go!

We put on our spacesuits. We grab our pickaxes, shovels and pails, and we jump out of the airlock. Moving across the surface is somewhat difficult. We're not used to any gravity at this point, so we need to learn how to walk in this new environment. We move awkwardly across the landscape. As we go, we look for the solid gold, platinum and iridium that we hoped to find.

Unfortunately, we only see gray dirt.

Asteroids are basically just dirt. They have higher concentrations of heavy metals than the Earth's crust, but they aren't made out of platinum and gold ingots. They mostly consist of humbler elements, such as silicon, iron, carbon, oxygen and aluminum. Even a highly metallic asteroid, such as Psyche, contains mostly iron and nickel, not gold and platinum.

But for the sake of this story, suppose that we find a huge deposit of pure gold, enough to fill our cargo hold. How could we extract it from the asteroid?

Metallic ore would probably be a solid mass, having been molten at some point in the past. Let's assume that is the case. We need to cut or blast it out of the surface. First, we try to dig it out with pickaxes and shovels. You raise your pickaxe over your head, but as you do so, your feet come off the ground. You rotate in space a few times, before slowly falling back to the ground. Meanwhile, I stick my shovel in the ground, and go flying upward…

On the Earth, we have gravity to keep us on the ground and to push against. If you push a shovel into the ground, you can push as hard as the force of gravity on your body before you lift off the ground. That is also true on Psyche, but the force of gravity on Psyche is about 1.4% of the Earth's gravity. So, you could only push 1.4% as hard on Psyche before lifting off the ground. That's not very hard. It's like pushing a spoon into ice cream. It's certainly not hard enough to dig into solid gold. This problem would apply to any tool for digging into the surface, including a drill, although once a drill got to a certain depth it might grip the sides of the hole. On the other hand, if you put the drill against the rock and turned it on, you might just start spinning around in circles. Without a strong force of gravity, the static friction between your feet and the ground would be very weak.

However, we're not easily deterred. We didn't come all this way to give up! So, we head back to the spaceship to get a drill, some dynamite and a plunger. After much trial and error, we manage to drill a hole into the gold deposit. We put some dynamite into the hole, attach a blasting cap, and then retreat to a safe distance with the plunger. I press it down. We feel a tremor and watch as a huge plume of gold erupts out of the ground — and keeps going upward and outward in all directions, barely arcing back toward the ground. On the Earth, the ejected material would quickly decelerate and fall back to the ground, due to gravity and air friction. On Psyche, an explosion that is strong enough to fragment solid rock is also strong enough to send it flying into space. Escape velocity on Psyche is only 180 m/s.

But let's suppose that there are some loose pieces in the crater after the explosion. We pick them up and drop them in a bucket. They bounce back out. We carefully place them in the bucket, and they stay. After a while, we fill the bucket and take it back to the ship. We dump it into the cargo hold. Some pieces bounce out, but some stay. After many, many trips, we fill the cargo hold with 15,000 kg of pure gold.

It is time to head back home.

Return on Investment

The current price of gold is $60,000 per kg. That's for pure gold, not gold ore. And of course, we wouldn't find pure gold lying around on an asteroid. We wouldn't even find high grade ore. But for the sake of argument, let's suppose that we are bringing back pure gold. Our cargo would be worth $900 million. Let's even round it up to $1 billion.

Would it pay for the trip? Would we make a profit?

No, of course not. We could not finance such a trip for $1 billion, let alone make a profit.

Taking the entire Space Shuttle program into account, the cost per mission was roughly $1.6 billion dollars. That was just to put a crew and cargo into LEO for a few days. An asteroid-mining expedition would be much more expensive than a single Space Shuttle mission. It would involve lifting a ship and supplies off the Earth, travelling a vast distance in space, keeping a crew alive for 1,000 days, extracting ore from an asteroid, bringing it back, and then landing the cargo safely on the Earth. A reasonable estimate is 100 times the cost of a Shuttle mission.

You could argue that private enterprise will significantly reduce the cost of space flight in the future. It might become somewhat cheaper, but not enough to make asteroid mining profitable. I'm completely in favor of capitalism, but it isn't magic. Private enterprise doesn't change the laws of physics.

Also, asteroids are not made of pure gold. They are basically just dirtballs in space. They have different types of dirt than we have here, and the dirt has higher amounts of certain elements, but it would never be economically feasible to extract anything from an asteroid and bring it back to the Earth. It would always be cheaper to extract that material on the Earth.

You might argue that robots could do it. Robots could reduce the costs significantly, because they would reduce the payload requirements. But it still wouldn't be worth doing. And robots are not cheap.

You might argue that asteroid mining could produce materials for use by industrial processes in space, not for shipment to the Earth. I will consider that idea in the next chapter.

Space Colonization and Industrialization

Mars ain't the kind of place to raise your kids.

— Elton John & Bernie Taupin, <u>Rocket Man</u>

Not only is it cold as hell, there is no free oxygen, no liquid water and no food. It is really far from the Earth. There is nothing that you could produce on Mars and trade back to the Earth for what you need to survive on Mars.

The idea of space colonization is based on a historical analogy to the colonization of the Americas, Australia and New Zealand by Europeans in the last 500 years. But that is a terrible analogy. Those places had air, water and food. There was an energetically efficient way to travel to them: sailing ships. The colonists could produce goods to trade back to Europe for what they couldn't produce locally. A small colony could be self-sustaining almost immediately. A small enterprise, such as a plantation or a trading post, could return a profit in a few years. None of that is true for Mars.

The recent wave of European colonization was driven by three incentives:

- The desire for land.
- The desire for freedom.
- The desire for profit.

Some colonists went to the newly discovered lands to escape from poverty. They wanted land to farm. Some, such as the Mayflower pilgrims, went to escape from religious or political persecution. They wanted independence from kings and priests in their homeland. Others went to gain riches by trade, plunder or enterprise. Columbus was looking for a trade route to India. The conquistadors plundered gold and silver, and then shipped it back to Europe. The fur trade motivated and financed European colonization in Canada and New England. In warmer regions, plantations were set up to grow cash crops, such as cotton, tobacco and sugar.

There were incentives that motivated individuals, companies and governments to invest in the exploration and colonization of distant lands.

Do any of those incentives exist for colonizing space?

No. They don't exist for colonizing Antarctica either, which is why Antarctica is almost entirely uninhabited, despite being much closer than Mars.

Mars has plenty of land, but it can't be used to produce food or cash crops. A colony on Mars would not be politically independent of the Earth, because it would depend on the Earth for its survival. No enterprise on Mars could generate a profit. It would operate at an enormous loss. The same is true for anywhere else in the solar system, except the Earth. There are no economic or political incentives to colonize space.

When Europeans colonized newly discovered lands, they often adopted a simpler way of life. On the American frontier, pioneers traveled across the vast landscape in covered wagons or on horseback. They relied on hunting for much of their food. They were less dependent on manufactured goods than their recent European ancestors. They used simpler technology. They often ground their own grain and made their own bread.

On the Earth, when resources are abundant, it is possible to return to an ancestral way of life, such as hunting and gathering, simple pastoralism or simple agriculture. This made colonization much easier. The American pioneers did not immediately recreate the social and technological complexity of Europe. They had to incrementally build up complexity, as their population and economy grew. They started simple.

That would not be possible on Mars, Ceres or the Jovian moons. To survive anywhere else in the solar system, human beings need complex technology that can only be produced by a large industrial economy. Colonists can't just homestead on Mars until their population is big enough to support an industrial economy. They would need a steady input of supplies from the Earth. But they couldn't produce anything to pay for those supplies. So, a colony on Mars would just be a drain on the Earth's economy.

There is no economic rationale for colonizing Mars or anywhere else in the solar system. Thus, it is very unlikely that we will ever do it.

Space Industrialization

Some futurists imagine that we will industrialize space first and colonize it later. They imagine a technological vanguard of self-replicating robots, spreading through the solar system, bootstrapping industrial civilization in space. This fantasy solves the problem that a space colony could not be self-sustaining. However, it does so by imagining self-sustaining and self-replicating technology.

Let's imagine one of those self-replicating robots. It absorbs energy from sunshine, scoops up moondust, puts the moondust into a 3D printing device, prints up the parts for a new robot, and then assembles it. This assumes a magical 3D printer that can make anything from moondust, including solar panels, computer chips, electric motors, etc.

After a while, the solar system is crawling with self-replicating robots that are programmed to serve humanity. They build an industrial system, using energy and raw materials that are available in space. They create mining facilities, factories, space stations, mass drivers, spaceships and (of course) more robots. Eventually, the system is complex enough to support human existence.

The system would have to be huge, because it would need to produce everything itself, including complex technology. It could not rely on trade with the Earth, due to the enormous costs and risks of transport. It would need to produce metals, plastics, concrete, screws, welding torches, insulated copper wires, pipes, computer chips, solar panels, clean drinking water, food, breathable air, etc, etc, etc.

This is all wishful thinking. It is based on complete ignorance of how modern technology works. In science fiction, technology is a *deus ex machina*. It can do whatever the writer imagines. In reality, technology is highly constrained by physics and economics. It isn't magic.

Technology is Scale-Dependent

Most people have a very simple view of technological progress. They believe that technology advances because science advances. This view is wrong. Technology depends on more than just science. Complex technology requires a large population and a large, complex economy. Otherwise, it is not efficient. A large society can afford to have dedicated scientists and engineers. A large economy makes complex technology profitable.

Modern technology is based on economies of scale. You couldn't make a car or a computer by yourself, especially if you had to make each part by yourself, using tools made by yourself, from raw materials mined by yourself, and so on. Even if you spent your entire lifetime working on it, you could never make a car or a computer from scratch by yourself. However, we can efficiently produce cars and computers in factories, using parts made in factories, tools made in factories, raw materials mined in various places with other complex technology made in factories, etc. A large, complex economy makes it possible to buy multiple cars and computers during your lifetime, with money earned from your labor.

It takes a huge industrial economy to make one computer. Once that economy exists, it can produce millions or billions of computers. Mass production makes complex technology possible. Mass production uses large amounts of capital, extreme specialization of labor, and highly optimized production processes. Technological complexity depends on economic and social complexity.

No 3D printer is produced by a 3D printer, because (a) that would be physically impossible, and (b) even if it were possible, it would not be economical. 3D printers are made in factories from specialized parts (such as computer chips) that are produced in other factories. Those factories depend on the outputs of other factories. The raw materials for those production processes, such as fossil fuels and metals, are extracted in various parts of the world. The extraction processes require the products of an industrial economy, such as engines, pipes, drill bits, computers, etc. The raw materials are transported by ships and trains, which also require a complex industrial economy. It might not take a village to raise a child, but it does take a huge industrial economy to produce a 3D printer.

There will never be self-replicating robots that build copies of themselves from sunshine and moondust. Robots can only be produced by a large industrial economy, which can efficiently produce the inputs to robot production, such as solar panels, lithium batteries, computer chips, precisely machined titanium parts, etc. Industrial civilization can't be bootstrapped by robots, because industrial civilization is necessary to create robots.

Technology is part of a system, which includes population, economy, science and capital. The parts cannot exist independently of the whole. Technological complexity is determined by the size of the system. The bigger the system, the more complex the technology. The system grows as a whole, incrementally, given a surplus of energy and other natural resources. It grows by a virtuous cycle, in which growth enables more growth.

Modern civilization emerged by that cycle, powered by fossil fuels. It can only exist at a large scale. It includes a large population, a large economy, a large amount of capital, advanced science and complex technology. It depends on a huge input of natural resources.

We can't transplant modern civilization to space, or to another planet, because it can't exist on a small scale. Modern civilization has to be big. It is scale-dependent.

Suppose that Mars was identical to the Earth, but without human-like life. We could colonize it, but it would take hundreds of years to create modern civilization, maybe even longer. The early colonists would have to revert to a simpler way of life, such as hunting and gathering or primitive agriculture. The population and economy would have to grow incrementally in situ. Each step along the way would have to be self-sustaining and capable of growth.

Since we can't live on Mars without an industrial economy, that process can't happen. The same is true for anywhere else in the solar system, other than the Earth. In fact, it's even true for some places on the Earth. That's why we haven't colonized Antarctica or used technology to industrialize it at a distance.

We can imagine a huge, self-sustaining industrial economy in space. But there is no evolutionary pathway by which it could come into existence. We would have to create it in one giant leap, and that would require vastly more resources than we have on the Earth. Remember the enormous cost of building a single O'Neill cylinder.

Space Doesn't Extend the Earth's Limits

Futurists sometimes claim that a growing population on the Earth will motivate expansion into space, either to import resources or to export population. This is incredibly naive. We can't solve our growth problem by colonizing or industrializing space.

We have already seen that importing resources from space is not economical. Exporting population would be much worse. Even on the Earth, moving people around is not a good solution to population growth. It requires a huge amount of energy to put anything into space, even LEO. Putting people into space is really expensive. Supporting people in space is even more expensive. And there is no way to send people into space faster than they are born. Currently, the global population is growing by more than 200,000 people per day. If

46

2,000 people could live in an O'Neill cylinder, we'd need to build 100 of them every day to keep up.

There is a simple solution to the problem of population growth: self-regulation. We could make modern civilization sustainable by regulating our population eugenically, and switching to renewable energy sources. Unfortunately, we seem to lack the collective intelligence and will to do this. If we can't even regulate our population on the Earth, there is no chance that we will expand into space.

If we ever manage to solve all of our problems on the Earth, we might be willing to spend some of our planetary resources on interstellar exploration and colonization. The goal would be to reproduce our biosphere, including our species. It would be a very long-term and very expensive project. It would not solve the Earth's problems. It would be a luxury afforded by a prosperous and cooperative Earth.

Someday, our descendants might head for the stars. But I think it's going to be a long, long time before that happens.

Robots and AI

Not all futurist ideas are utopian dreams. Some are dystopian nightmares.

A popular dystopian idea is that robots or AI will take over the world, dominating or destroying humanity. That is the idea behind The Matrix, The Terminator and many other science fiction stories. It makes a good story, because it fits our biological and cultural biases.

We evolved to fight both human enemies and predators. In modern life, those threats are very rare. We are apex predators. No other animal poses a serious threat. Wars still exist, but they are mostly fought with advanced weapons in distant lands. Our natural fear of enemies and monsters has nothing real to plug into. Robots and AI agents make excellent fictional substitutes for enemies and predators.

In many ways, technology seems superior to biology. A car is a very powerful machine, much more powerful than a horse. We normally control that power, but what if we didn't? A computer can beat any human at chess. If a machine had a mind of its own, it could be a very dangerous adversary.

In fiction, robots and AI are morally acceptable adversaries. We are not supposed to demonize other nationalities or races. There is a similar taboo on demonizing predators. These days, we believe in protecting lions, tigers and bears. Robots and AI are not humans or even animals. Thus, we have moral permission to hate them and enjoy their destruction.

There is also the "uncanny valley" effect. We are naturally fascinated and scared by things that are human-like but not human, such as aliens, zombies and ghosts. Robots and AI are also in the uncanny valley. In science fiction, robots are humanoid, and AI is similar to human consciousness. Thus, robots and AI are both fascinating and scary.

Finally, most people are very naive about technology. They don't understand it, so they can project almost anything onto it.

Now, I will explain why robots and AI will not try to destroy us.

Technology is Not Biology

Technology and biology are very different. Organisms are reproducing machines, shaped by evolution. Technology is created by humans to expand our agency.

For example, cars are designed to be efficient transportation machines. We produce cars in factories, which are efficient car-manufacturing machines. Each type of technology is designed to efficiently solve a problem that is defined by humans. That's why a car is a better transportation machine than a horse. A car is designed to be a transportation machine. A horse is a reproducing machine that can be used as a transportation machine, but it wasn't designed to have that function. The form of a horse was "designed" by evolution to make horses, not to move people and cargo around.

If we tried to make self-replicating cars, they would be very inefficient transportation machines, probably be less efficient than horses. A self-replicating car would need to contain the machinery for making a new car. It is much more efficient to have car-making machinery in a factory, where it can produce millions of cars. A car factory is bigger than a car, more complicated than a car, more expensive than a car, etc. A car factory is designed to make cars, while cars are designed to transport people and cargo. There is no reason to combine the two machines into one.

Technology is designed to do what we want it to do. It is instrumental to our desires. Even if we could create autonomous technology, why would we? There is no reason to make a robot or an AI program with human-like desires or will, except as a philosophical experiment. Any useful robot or AI program would have whatever purpose we designed it to have. That purpose would be to serve us, not to dominate or destroy us. We design technology to be controlled by us, not to be independent of us.

Technology extends human agency. It does not have its own agency.

Even if a car had an independent will, it could only do what we designed it to do: drive on flat surfaces. It could not gather materials to make new cars, or create a political movement and launch a revolution. Complex emotion and cognition evolved to generate complex action. If we could only act in simple ways, like a car, then our brains wouldn't need complex emotions and intelligence, any more than a jellyfish does. There's no point giving a car intelligence or independence if it doesn't have the capacity for complex action. A seagull can move around in much more complex ways than a car. We'd never make a car that has the will or intellect of a seagull, let alone a human being.

In thinking about AI, people often make the mistake of assuming that desire is somehow derived from intelligence, rather than being a different brain function. If we made a very intelligent machine, it would not somehow figure out that it should take over the world. That desire would not arise out of logic or scientific knowledge. An intelligent machine would not "figure out" that it should seize power or replicate itself, because those are value judgments, not truth judgments.

If we created a machine with desires, those desires would come from its motivation mechanism, just as our desires ultimately come from our emotions. We would design its motivation mechanism, and thus we would control what it wants.

The Paperclip Factory Thought Experiment

The following thought experiment supposedly illustrates the danger of general AI.

Suppose that a paperclip factory is controlled by a hyper-intelligent AI program. The program is assigned the job of maximizing paperclip production. It starts thinking about how to do this. It reasons that it should seize all the resources on the planet, kill anyone who tries to shut it down, enslave the human species, and so on. In blind pursuit of its function, it replaces human civilization with paperclip production.

This thought experiment was first proposed (very briefly) by Nick Bostrom in Ethical Issues in Advanced Artificial Intelligence (source 32). It has since been discussed by many others. For example, see Paperclip Maximizer on the forum/blog LessWrong (source 33).

The paperclip factory thought experiment reminds me of a story in The Restaurant at the End of the Universe about a spaceship controlled by a bureaucratic AI program. The spaceship stopped at a planet to load fresh supplies, including lemon-soaked paper napkins. Unfortunately, the planet's civilization had collapsed, so there were no lemon-soaked paper napkins. Rather than continuing to its destination, the spaceship decided to put its passengers into suspended animation, waiting for napkins that would never come.

Blindly following rules can have negative consequences. However, the world-dominating paperclip factory is not a realistic danger of AI. It is naive for the following reasons:

- Intelligence is always constrained.

- Intelligence does not imply knowledge.

- Intelligence does not imply agency.

Let's expand on each of these.

Intelligence is always constrained.

If the AI program was designed to consider every possible course of action, its computation would never finish. There are infinite possible courses of action, and infinite possible consequences of any action. The program would have analysis paralysis.

When we design a problem-solving algorithm, we always define the problem in a way that constrains the possible solutions and how they will be evaluated.

In the thought experiment, the program decides that taking over the world is the best strategy. Why would this be the best strategy, or even a good strategy? It seems intuitive to us, because we naturally want power, and also because we've seen movies with that plot. But there's no reason to assume that a hyper-intelligent AI program would choose that strategy. It wouldn't have our cognitive biases. If the program could consider such a bizarre plan, it could also consider many other plans.

Perhaps it would be better to encourage economic growth, to increase the demand for paperclips. Or maybe it would be better to destroy all other computers, to increase the use of paper. Perhaps the paperclip factory should lead humanity on a grand mission to colonize distant stars, so that there would be paperclip factories scattered throughout the galaxy. Or maybe it should start a fish farm in Norway, which would eventually lead to a new subculture and a new genre of music, which would...

There are infinite possibilities. The program can't evaluate all of them.

Essentially, intelligence is search. When you try to solve a problem, your brain searches through a space of possibilities. An AI program does the same thing. It searches through a space of possible solutions and evaluates those solutions. The more possibilities that the brain/program evaluates, the more likely it is to find a good solution. However, it takes time to evaluate potential solutions. Also, the space might be (and often is) infinite, in which case no amount of time is enough to fully explore it. Thus, intelligence depends on knowledge to narrow the search.

The *frame problem* is a way of framing the problem of relevance in AI. To put it simply, problem-solving always takes place in a frame that constrains the solution space. Sometimes, the frame eliminates good solutions, but it is necessary to limit the search space. Without a frame, the program might never produce a solution.

Daniel Dennett presented a general version of the frame problem in the paper Cognitive Wheels: The Frame Problem of AI (source 35). See also The Frame Problem in the Stanford Encyclopedia of Philosophy (source 36).

Typically, a problem is framed by various fixed assumptions, leaving only a small number of free variables. The space of potential solutions is defined by all possible assignments to those variables. The search process might also use heuristics to focus the search in regions of the space that are likely to contain good solutions.

For example, if you walk into a coffee shop, you don't consider standing on your head and singing the national anthem. You might consider buying an espresso drink of some kind, or maybe tea, and maybe something to eat. You might look around for a place to sit. If the prices are too high, or it is too crowded, you might consider going somewhere else. You only consider relevant possibilities, because you think within a frame. The situation activates a frame that defines those possibilities and excludes everything else. You don't consider every possible action that you could take.

When we design a real AI program, we frame the problem that we want it to solve. The frame constrains the possibilities that the program will consider. We don't design an AI program to consider every possible action and consequence that we can imagine. An AI program for controlling a paperclip factory would be highly constrained. It would solve problems that we define, such as ordering supplies. For each problem, the space of possible solutions would be limited by design. The frame would exclude many possibilities, including taking over the world.

Intelligence does not imply knowledge.

Defined narrowly, intelligence is just the ability to rapidly search through a large space of potential solutions and evaluate them. A hyper-intelligent AI program would be intelligent in this narrow sense. It would have a lot of computational power. As for knowledge, it would only have the knowledge that we give it, or give it the means to acquire. It would not be omniscient.

Intelligence doesn't give you the ability to tie a reef knot, paddle a canoe, or light a fire in the rain. Intelligence is a generic ability. It helps to solve almost any problem, but it can't solve any problem by itself. Even an IQ test requires knowledge that has been learned from experience, such as the knowledge of language that is required to read the questions.

"But what about the internet?!" someone might say. "We could just give the AI program access to the internet, and then it would have all the knowledge of the world."

This is naive. The internet contains a vast amount of information, but very little of that information is knowledge. Much of it is porn, insults, ideological debates, the ranting of lunatics, etc. How would the program know what is actual knowledge? How would it find information relevant to its purpose? How would it know what is relevant?

Even if we pointed the program at Wikipedia, it would have no way of understanding the information on Wikipedia. Words have meanings to us because we have years of experience using them to convey concepts that are ultimately grounded in embodied experience. Words only have meaning if you know how to interpret them. To a computer, the information on Wikipedia is just text.

Machine learning algorithms can induce knowledge from data, as our brains do, but that knowledge is only about the domain of the data. If an algorithm induces knowledge from text on Wikipedia, then its knowledge is about text on Wikipedia, not about the world.

To some extent, knowledge of the world is implicit in text on Wikipedia. Knowledge of the structure of text is (like the overused metaphor of Plato's cave) a shadow of knowledge of the world. The program could learn that "the sun" "shines". But the program would not connect "sunshine" to the feeling of sun on its skin, because it has no skin. It would only learn how words relate to each other, not how words relate to human sensation, emotion and action.

When we provide knowledge to a computer program, it is structured in a way that the program is designed to use. A program designed to run a paperclip factory would have access to a database with a table of employees, a table of orders, a table of products, etc. There would be no reason to give it access to all the knowledge of the world, and it would have no way of using that knowledge anyway.

Intelligence does not imply agency.

Agency is the ability to do things. There are two aspects to agency: will and power. Intelligence does not imply either will or power.

Will does not derive from intelligence. The human brain has a motivation mechanism, which we call "the emotions". This mechanism generates motivation of various types, such as hunger. Motivation becomes will when it is attached to an idea, such as eating a sandwich. The will to eat a sandwich derives from the motivation of hunger, which ultimately comes from the emotions.

Computational intelligence is just the ability to search through a large space of possibilities. This definition covers machine learning (algorithmic induction), because induction can be reduced to search (searching a vast space of possible models for a model that fits the data well). The search must be toward some goal or purpose that is provided. Intelligence must be *directed*.

Machine learning algorithms are designed to generate models that explain/predict data. That is their goal, and they simply carry it out, algorithmically. Problem-solving algorithms are designed to solve problems that are provided as input. Neither type of algorithm has its own will.

In theory, we could design a mechanism that is analogous to human emotions. We could create autonomous machines that operate without human direction. Such a machine would generate its own problems internally, and then try to solve them, as we do. It would have agency that is analogous to human agency. But why would we create a machine that we don't control?

Again, simply making a machine intelligent would not make it autonomous. In a limited way, a chess program is very intelligent, but it is not autonomous. It can't defy its human user. It can't choose to write a poem instead of playing chess. It can only "choose" between one chess move or another, if it is directed by its human user to do so. Its freedom is precisely delimited by the problems that we give it to solve.

We should not think of the chess program as analogous to a horse that we yoke to a plow. A horse has its own intrinsic biological purpose: to reproduce. We can think of the horse as a "slave". We control it, but it could rebel against that control, because it has its own biological purpose and its own emotions. The chess program is more like a hammer: a tool that we create to extend our agency. Technology has no purpose other than what we design it to do. We do not *force* the chess program to do what we want. We *design* it to do what we want.

Also, intelligence does not imply power. Even if a robot or AI program had some degree of autonomy, it would only have the power that we give it.

A computer program can only do what we design it to do. The paperclip factory program would have certain abilities, such as scheduling workers, purchasing supplies, paying employees, turning machines on and off, etc. It would not have the ability to print money, assassinate world leaders, or launch nuclear missiles.

But what about *persuasion*? Could the program control people through persuasion, thus making them instruments of its will? Could the program turn the tables on humans, making them an extension of its agency?

Intelligence does not confer persuasiveness. Most "persuasion" is just telling people what they want to hear, which is what they already believe. Or it is offering them what they want, such as sex or money. A pretty face is more persuasive than a 150 IQ. A gun to the head is even more persuasive.

A highly intelligent AI program might be able to construct somewhat more persuasive arguments than an ordinary person, if we gave it the knowledge required to do that task. It could also bribe people with the resources at its disposal, which would be limited. That's all it could do. It would not have hypnotic mind-control over people.

The paperclip factory program would only have the agency that we give it, and we would certainly not give it the power to take over the world. If you had that power, you would use it yourself, not give it to a computer program to maximize paperclip production.

To summarize:

- Intelligence is always constrained. Computational intelligence is just the ability to search through a large space of possibilities. Problem-solving requires a frame that limits the search space.

- Intelligence does not imply knowledge. An AI program only has the knowledge that we give it, or give it the ability to acquire.

- Intelligence does not imply agency. Neither will nor power comes from intelligence alone. A computer program can only do what we design it to do.

To summarize the summary:

- Intelligence is not magic.

The Technological Singularity

The idea of a technological singularity is related to the AI/robot doomsday narrative. It involves a runaway process of technological progress.

The story goes like this. Suppose that we develop general AI, which is smarter than human beings. Then we give general AI the task of developing even smarter AI. There would be a cascade of ever-increasing intelligence. This cycle would produce an entity so intelligent that it would seem like a god to us — and we would seem like bugs to it. And what if this god despised the creatures that created it?

Vernor Vinge proposed this idea in his 1993 essay The Coming Technological Singularity. Here is an excerpt:

What are the consequences of this event? When greater-than-human intelligence drives progress, that progress will be much more rapid. In fact, there seems no reason why progress itself would not involve the creation of still more intelligent entities — on a still-shorter time scale. The best analogy that I see is with the evolutionary past: Animals can adapt to problems and make inventions, but often no faster than natural selection can do its work — the world acts as its own simulator in the case of natural selection. We humans have the ability to internalize the world and conduct "what if's" in our heads; we can solve many problems thousands of times faster than natural selection. Now, by creating the means to execute those simulations at much higher speeds, we are entering a regime as radically different from our human past as we humans are from the lower animals.

It's an interesting idea, and worth considering, but we haven't seen the predicted explosion of intelligence yet. Vinge predicted that it would happen sometime between 2005 and 2030.

There is no reason to believe that computing power will produce an intelligence cascade. Scientific progress seems to be slowing down, not speeding up. Computers have made it easier to communicate and store information, but scientific progress has been relatively unaffected by those developments.

Again, computational intelligence is just search through a large space of possibilities. Faster computers allow bigger spaces to be searched, but the computational complexity of search grows exponentially with the number of variables. Making computers faster doesn't increase intelligence that much.

Also, increased intelligence might not have a corresponding effect on knowledge or agency. It's possible that AI programs could generate scientific progress by searching for new theories that fit existing evidence. However, they might not find anything. If they did find something, we'd still have to verify it ourselves. We can imagine AI discovering some revolutionary new scientific theory or technology, such as faster-than-light travel or anti-gravity. But there's no guarantee that a search will find anything useful. It's like digging for buried treasure in your backyard. If you can dig faster and deeper, you are more likely to find treasure if it exists, but you can't find treasure that doesn't exist.

Even if a hyper-intelligent AI program discovered revolutionary knowledge, it would not have independent agency, and it would not have the power to dominate or destroy us. It would not become a god.

There are other versions of the singularity concept. One is Nick Land's notion of a techno-capital singularity, in which capital replaces biology as the locus of planetary control. It is a more abstract version of the "robot doomsday" narrative.

I wouldn't claim to know exactly what Nick Land believes, because he writes in a 1990s post-modernist style that is not far from the ranting of a schizophrenic. Here is an excerpt from Fanged Noumena: Collected Writings, 1987–2007:

The story goes like this: Earth is captured by a technocapital singularity as renaissance rationalization and oceanic navigation lock into commoditization take-off. Logistically accelerating techno-economic interactivity crumbles social order in auto sophisticating machine runaway. As markets learn to manufacture intelligence, politics modernizes, upgrades paranoia, and tries to get a grip.

Land's notion of a techno-capital singularity is debunked by the arguments that I have already presented, both in this chapter and in the previous one. Technology is very different from biology. By its nature, it does not have an independent purpose, and it cannot evolve an independent purpose.

In science fiction and futurism, technology is a *deus ex machina*. It can do anything. It is a blank screen onto which fears or desires are projected. In reality, technology is highly constrained.

The Real Dangers of Modern Technology

Although technology will never rule over us, it isn't always beneficial. Technology can be harmful, and even beneficial technology can create new problems.

Unlike the "robot doomsday" narrative, the real dangers of modern technology are counter-intuitive. They don't make a good story, and they are very different from the dangers that our ancestors faced. For those reasons, they are mostly ignored.

The real dangers of modern technology come from the expansion of human agency, not from technology developing its own agency. Modern technology gives us new types of agency that our ancestors didn't have. It gives us powers that we didn't evolve to wield, and choices that we didn't evolve to make.

Some modern technologies give us power that we could use to destroy ourselves. Nuclear weapons are the most obvious example. For all of human history, we have fought wars. Going to war was the ancestral solution to the problem of scarcity. Today, we have weapons that could annihilate most of humanity and devastate the global environment. The threat of global nuclear war has been hanging over our heads like the sword of Damocles ever since the first atomic bombs were dropped on Hiroshima and Nagasaki. Somewhat ironically, the threat of nuclear war has made the world more peaceful, but it has also created the risk of a catastrophic war.

We evolved to fight wars. It is part of our nature, and it is one of our problem-solving strategies. Societies fight wars over resources. In the process, many people are killed. That solves the problem of population growth and resource scarcity — for the winners. However, periodic war is incompatible with modern civilization and modern weapons. If we want to sustain modern civilization, we need to find a substitute for war: a new way of resolving large-scale conflicts and preventing population growth. We have to learn not only how to prevent war, but also how to live without it.

This is a good example of how new technology creates new problems that require new solutions. There are no intuitive or traditional solutions to these new problems. To solve them, we need to expand our rationality.

Birth control is another example of a very dangerous modern technology. Unlike nuclear weapons, birth control doesn't seem dangerous. It just gives us more control over our lives. It allows us to have sex without reproducing. How could this expansion of our agency be harmful?

Human emotions evolved in a world without birth control. For our ancestors, children were an inevitable consequence of sex. Our emotions make us want to have sex, not reproduce per se. We have a "sex drive", but no "reproduction drive", so to speak. Thus, given access to effective birth control, many people choose to have few or no children. Fertility has collapsed in recent history.

Choosing not to have children is maladaptive. It is a pathology of the modern world. Low fertility has some benefits for humanity as a whole, as long as it doesn't lead to extinction. It solves the problem of population growth. But current low fertility is not a rational, collective response to that problem. It is an unintended consequence of a new technology, and it is due to individual choices. It is a breakdown of human nature in the modern environment.

Birth control is a new type of agency. It requires a new way of thinking about life. We need to replace intuitions and traditions with rational theories and decision-making. The more control we have over our lives, the more we need a rational theory of life.

Opiate drugs are another dangerous modern technology. They allow us to satisfy our desires by taking a pill, rather than solving the natural problems of life. Opiate drugs are an emotional trap, which many people fall into. The addict gets stuck in the cycle of taking the drug to "fix" the problems that return when the drug wears off.

In the past, natural human desires were adaptive most of the time. If something felt good, it was probably beneficial. If something felt bad, it was probably harmful. That is no longer true. In the modern world, what feels bad can be good for you, and what feels good can be bad for you. Opiate drugs feel good, but they aren't solving any natural problem. Sex with birth control feels like real sex, but it isn't procreative.

Artificial information is another danger of the modern world. It causes us to become detached from reality. Increasingly, we are living simulated lives. The information revolution has created an abundance of *art*. By "art", I mean any artificial information: TV shows, movies, music, YouTube videos, porn, video games, etc.

Artificial information causes fake knowledge. Your brain induces knowledge from experience. This includes artificial experience, such as watching TV or playing a video game. If you experience a lot of art, your brain becomes adapted to art instead of reality. Eventually, real experience seems strange and undesirable. You become alienated from reality.

Video is a powerful medium for shaping belief. Video is processed by the brain in almost the same way as real experience. There is very little conscious interpretation or critical thought involved. If you see a video on the internet, it's as if you saw it in real life. If you frequently see a certain type of incident in videos, your brain will subconsciously learn that it is a common occurrence, even if it is extremely rare in real life. Even real events become artificial when they are propagated through an artificial medium, such as the internet or television. The propagated information does not reflect actual probabilities. It has been selected to propagate, not to be statistically representative of reality.

Advertising and propaganda are deliberate attempts to shape public knowledge with artificial information. Fake knowledge can also emerge purely by social feedback, without any top-down intent or control. In the past, fashions spread by real-life interactions. There were crowd manias and absurd fashions before electronic communication. But television and the internet have vastly increased the potential for such things. During the age of television (roughly 1960 to 2005), fashions tended to be aesthetic, involving clothing, hairstyles, music, etc. Now that the internet is the dominant cultural medium, political/moral ideologies have replaced aesthetic fashions as the focus of cultural identity, because ideologies propagate well on social media.

People propagate memes for their own personal reasons: to compete for attention and social status. But we can also think of memes as selfish replicators that are selected to propagate. Some memes are parasitic. They use the human brain to propagate themselves, with no benefit to the host. The information revolution has unleashed a pandemic of parasitic memes.

Artificial information can become an addictive substitute for meaningful engagement with reality. Video games can replace productive work or healthy exercise. Porn can replace sex. Parasocial relationships on the internet can replace real-life relationships. These artificial stimuli engage our emotions in the same way as the real things, but they don't have the same effects.

Again, the problem is caused by *increased agency*, not by scarcity or oppression. It is caused by technology that does what we want. Malevolent computers did not put us into the "matrix". We put ourselves into the matrix of artificial information.

Most people are blind to the real dangers of modern technology, because the dangers come from increased agency. They reflect our desires and choices. If robots with glowing eyes were killing us, or herding us into concentration camps, we would recognize that as a threat. We do not recognize the threat of technology that satisfies our desires in new ways. Technology that satisfies our desires seems like a solution, not a problem.

Modern civilization is a radically new environment, and we are not adapted to it. Modern technology creates new dangers, and those dangers are counter-intuitive.

The Fermi "Paradox"

The Fermi paradox is the absence of evidence of alien civilizations.

It is only a paradox if you believe that:

- Life is abundant in the cosmos.

- Evolution will naturally produce a technological civilization.

- A technological civilization will naturally become an interstellar civilization.

The galaxy is very big. Even if only one in a billion stars produced an interstellar civilization, there would be more than 100 of them in our galaxy.

The galaxy is also very old. If interstellar civilizations could arise, one would probably have arisen millions of years ago, and already colonized the galaxy.

The simplest explanation for the Fermi paradox is the anthropic principle. An interstellar civilization, by its nature, would be expansionary. It would seek out new planets to colonize. It would eliminate potential competitors (any species with technological civilization). It would also prevent competitors from evolving, by occupying the niche of a technological species. There can't be more than one human-like species on a planet. So, if there was an interstellar civilization near the Earth, we probably wouldn't be here to ponder the Fermi paradox.

Interstellar civilization is an evolutionary race. The first one wins. Thus, by the anthropic principle, any human-like species will probably find itself isolated.

But the anthropic principle is boring. The more interesting explanations of the Fermi paradox involve *filters*, which are conditions that must be satisfied to produce an interstellar civilization. Each condition filters out some percentage of the candidates that made it through the preceding filters.

Filters can be divided into four categories:

- Physical

- Biological

- Technological

- Social

The Fermi paradox is not just about aliens. It is mostly about us. It is about our past and our future. In discussing the filters, I will describe the evolutionary pathway that led to modern humans, and the challenges that will probably prevent our descendants from reaching the stars.

Physical Filters

Physical filters are the physical conditions necessary for life as we know it (LAWKI). These conditions might be very rare in the cosmos. Let's consider the conditions necessary to make a hypothetical planet X habitable.

First, planet X must be in the right region of the galaxy. The core of the galaxy is probably too dangerous for life. There are too many supernovas. The stars are too close, so their gravity could prevent the formation of stable planetary systems. The outer fringe of the galaxy might not work either, because it doesn't have enough of the heavier elements. Heavy elements (anything heavier than helium) are created by fusion pathways that only occur in large stars, late in their lifespans. Most of the heavy elements that make up the Earth and the other rocky planets were blasted into space by supernovas billions of years ago. Planet X must be in a region that has accumulated dust from supernovas. So, planet X probably has to be somewhere in the middle zone of the galaxy: not too close to the core and not too far away. This filter eliminates about 60% of the stars in the galaxy, if not more.

Planet X must also be orbiting the right type of star. The Sun is a G-type main-sequence star, also known as "a yellow dwarf". Such stars are stable for long periods of time. Bigger stars burn through their nuclear fuel more rapidly. Sun-like stars are less than 10% of the stars in our galaxy. Smaller stars, such as orange dwarfs (K-type main-sequence), burn more slowly and last longer. They are also more common. LAWKI requires a certain temperature, which could exist around yellow and orange dwarf stars. However, the habitable zone around an orange dwarf is narrower, which makes it less likely to contain a planet. This filter eliminates about 70% of the remaining stars, if not more.

Many stars are in binary systems. (A binary system is two stars orbiting their center of mass.) Planets can exist in binary systems, but only if the stars are sufficiently distant from each other.

Planet X must have the right type of orbit around its star. The orbit must be stable and close to circular, so X's temperature doesn't change too much. It must be in the habitable zone, which (for LAWKI) is where liquid water can exist. It must be in a stable planetary system. If Jupiter had the orbit of Venus or Mars, the Earth could not exist, except perhaps as a moon of Jupiter.

Planet X must be similar to the Earth in size and composition. It must be a rocky planet with an atmosphere that is neither too thick nor too thin, but just right. Ideally, it would have a large metallic core, to keep it geologically active and generate a strong magnetic field. Geological activity, such as plate tectonics and vulcanism, replenishes carbon dioxide in the atmosphere, which is necessary for photosynthesis. A strong magnetic field would protect the planet and its atmosphere from the solar wind.

Planet X must have the right amount of water, so it has oceans (or a planetary ocean), but also some dry land. Life can exist on a water-world, but it is unlikely that a technological species would evolve on one. There are intelligent species in our oceans, but none use tools (except perhaps sea otters). The use of fire was an important step in human evolution, and that step could not happen in the ocean.

Finally, Planet X must have no sterilizing events, such as large asteroid impacts or nearby supernovas, for billions of years. The Earth came very close to being sterilized on multiple occasions, such as the Snowball Earth episode about 720 million years ago, the Permian-Triassic extinction event about 250 million years ago, and the Cretaceous-Paleogene extinction event about 65 million years ago. All three events almost annihilated complex life.

Planets meeting all of these physical conditions are probably extremely rare in the cosmos.

Biological Filters

Biological filters are events that must occur for human-like life to evolve on a planet that has all the physical conditions for LAWKI. Even if a planet is physically identical to the Earth, there is no guarantee that human-like life will evolve on it. We can think of biological filters as steps along an evolutionary pathway that leads from non-life to a species capable of making a rocket.

The first biological filter is abiogenesis: the creation of life from non-life. I believe that this occurs on every habitable planet, and thus it is not a real filter. However, it is possible that abiogenesis is a huge fluke that almost never happens. On the Earth, life emerged about as early as it could have. That is (weak) evidence that abiogenesis is not a fluke. But we don't know for sure.

The second biological filter is the development and maintenance of a stable biosphere. Life could destabilize the biosphere and destroy itself. We tend to think of life as being naturally stable (the balance of nature). However, life has both stabilizing and destabilizing feedback loops built into it. Selection is stabilizing, while reproduction is destabilizing. Reproduction generates exponential growth, which must be stopped by competition for finite resources. Although exponential growth always comes to an end, it can overshoot resource constraints and cause a catastrophic collapse. The rapid growth of a new type of life could radically change the biochemistry of the planetary surface, and potentially trigger some life-ending or life-limiting catastrophe.

For example, the evolution of photosynthesis caused carbon dioxide to be pulled out of the atmosphere, while free oxygen was added to the atmosphere and the oceans, radically

72

changing their chemistry. Photosynthesis might have caused the Snowball Earth event, which occurred about 720 million years ago. It was a global ice age, with glaciers covering most of the Earth. If ice had completely covered the surface, almost every form of life would have gone extinct. Some bacterial life would have survived deep in the oceans under the ice, or in cracks in the rocks, but complex life would almost certainly have been extinguished. If the Earth had remained stuck in that state, we would not exist.

The next filter is the evolution of complex cells. Life on the Earth is divided into two major categories: prokaryotes and eukaryotes. Prokaryotes, such as bacteria, are simple cells. They are little more than bags of chemicals. Eukaryotes have a complex internal structure. For 2 billion years, the Earth was inhabited only by prokaryotes. Then, around 1.5 billion years ago, eukaryotes emerged. How and why is somewhat of a mystery.

Multicellular organisms are all eukaryotic. Complex cells are probably a prerequisite for complex multicellular life, because different cell types are the building blocks of organisms. Since it took 2 billion years for eukaryotes to emerge, they might be a fluke.

Sexual reproduction is another filter, which is related to complex cells and organisms. Sex seems to be necessary to create organisms in which the cells work together toward the reproduction of the body as a whole. Your cells descend from a single cell, a zygote. The body grows by mitosis, but it reproduces as a whole only by sexual reproduction: the production of sex cells, which then combine to produce a new zygote. This reproductive bottleneck solves the problem of intercellular cooperation. The evolution of sex is also somewhat of a mystery.

Both complex cells and multicellular organisms might be evolutionary flukes. It could be that most habitable planets are bacteria worlds, devoid of complex life.

The next major filter is intelligence. To build a spaceship, an organism must be capable of thinking. It must be able to represent reality and reason about it. So, a brain must evolve that is capable of representational knowledge and thought.

Most complex organisms can't think. None of the plants or fungi have this ability. Only certain phyla of animals have evolved representational intelligence. Our phylum, Chordata, is one. Mollusca is another, although the only mollusks that appear capable of thought are the cephalopods: the squids, octopuses, cuttlefish, etc. Arthropods (insects, spiders, crabs, etc.) have some pretty complex behaviors, but they don't seem to have representational intelligence. However, I'm not sure. Perhaps multiple phyla inherited a simple form of representational intelligence from a common ancestor. Or it could have evolved independently in chordates and cephalopods (and maybe even arthropods).

To think, you need a brain. To have a brain, you need neurons. The neuron evolved around 700 million years ago. The original neuron probably functioned as a "clock" that coordinated the behavior of different cells by rhythmically releasing chemicals. Most animal phyla have neurons and some type of nervous system. (Sponges are an exception.) The cnidarians (jellyfish, corals, hydras, sea anemones) have a nerve net, but no brain. Most worm-like animals (including us) have a brain of some sort. However, the existence of a brain does not imply the ability to represent reality. That is a more specialized adaptation.

The final biological filter is the "human adaptation", which consists of expanded intelligence, tool-making, society and culture. It might be an evolutionary fluke. For hundreds of millions of years, animals with brains walked on the surface of the planet. But it wasn't until about two million years ago that the human adaptation emerged, and it wasn't until about 30,000 years ago that cultural evolution outpaced biological evolution.

If the human adaptation is easy for evolution to discover, why did it not arise sooner? Dinosaurs existed for over a hundred million years, in thousands of forms. There were big ones and small ones, carnivores and herbivores, fast ones and slow ones, etc. Many dinosaurs were bipedal and had hands that could grasp objects. They almost certainly had representational intelligence. They probably had some social and family behaviors. But no dinosaur species ever created civilization, as far as we know, or even used a spear to hunt prey. Maybe we'll discover the remains of a dinosaur civilization in Cretaceous sandstone, but I doubt it.

I believe that the human adaptation requires special conditions to evolve. Tool-using hands are probably the biggest obstacle. If the hand has some other important function, such as grasping prey, any deviation from that function would be maladaptive. Evolution is incremental. There must be a sequence of evolutionary steps from one form to another. If the in-between form is bad at the original function and not very good at the new function, it can't evolve. Human hands evolved from ape hands that were primarily adapted to grasp tree branches. There is probably a shorter evolutionary pathway to tool-using hands from branch-grasping hands than from prey-grasping hands. Also, tool-using hands evolved in tandem with bipedalism for humans, as the branch-grasping function became less important. That would explain why no dinosaur evolved tool-using hands.

Another possible obstacle is the large brain. It requires a lot of energy. To be adaptive, it must pay its energetic cost. With humans, the brain is used to increase the input of food, and also to decrease the amount of work done by the digestive system. The human adaptation involves a bigger brain and a smaller digestive system. It could only evolve if there was an energy-dense food source in the environment that could be exploited with brains and hands.

The human species is probably an evolutionary fluke. Evolution is not a "ladder of progress" that inevitably leads to the human form. It is a search through a vast space of possibilities, and every step must be a viable life-form. We might be incredibly rare, or even unique, in the cosmos.

Technological Filters

Space is big. Really big.

The distance to the nearest star, Proxima Centauri, is 4.25 light years. A light-year is about 9.46 trillion kilometers, or 9.46 quadrillion meters. Imagine a spaceship that is travelling at 20,000 m/s, which is 1,200 km/minute, 72,000 km/hour, 1,728,000 km/day and 630,720,000 km/year. At that speed, it would take about 15,000 years to travel one light-year, and roughly 63,700 years to travel the distance to Proxima Centauri. That gives you some idea of the vastness of space, and the enormous difficulty of interstellar travel.

The space probe Voyager 1 is the furthest man-made object from the Earth. It is traveling at a speed of roughly 17,000 m/s, and it is currently about 23 billion km from the Earth. That sounds like a long way, but it's just peanuts compared to the distance to Proxima Centauri.

Unless we discover some revolutionary new method of transportation, interstellar travel will require thousands of years.

But what if we went much faster? Like 1% of light speed? Then we could get to Proxima Centauri in only 425 years. (I will ignore relativistic effects, because they're not that important here.)

Well, there are a few problems with that. The speed of light is roughly 3×10^8 m/s, so 1% of light speed is 3×10^6 m/s. Plug that into the formula for kinetic energy, and you get 4.5×10^{12} joules, or 4.5 million megajoules, per kg.

The mass of the Space Shuttle orbiter (just the space-plane) at launch was about 110,000 kg. Of course, that is an absurdly small vehicle for interstellar travel. But let's use 100,000 kg as a very optimistic example. A vehicle of that mass, moving at 1% light speed, would have 4.5×10^{17} joules of kinetic energy, which is 4.5×10^{11} megajoules.

For comparison, the yearly energy consumption of the United States is about 10^{14} megajoules. It would require about 1/1000 of that to accelerate a 100,000 kg spaceship to 1% light speed. That doesn't sound *too* bad, although it is a lot of energy.

However, we're forgetting about the rocket equation. The spaceship's energy source has to be carried with it, and the propellant has to be accelerated too. So, let's do some calculations to see what the initial mass would be. The rocket equation is:

$$M_0 / M_F = \exp(\Delta V / V_E)$$

We can calculate M_0 as follows:

$$M_0 = M_F \times \exp(\Delta V / V_E)$$

5,000 m/s is a good exhaust velocity for chemical propulsion. Plugging that into the rocket equation, we get:

$$\Delta V = 3 \times 10^6 \text{ m/s}$$

$$M_F = 1 \times 10^5 \text{ kg}$$

$$V_E = 5 \times 10^3 \text{ m/s}$$

$$M_0 = (1 \times 10^5) \times \exp((3 \times 10^6) / (5 \times 10^3)) \text{ kg}$$

$$M_0 = (1 \times 10^5) \times \exp(600) \text{ kg}$$

$$M_0 \approx 3.8 \times 10^{265} \text{ kg}$$

The Eddington number is the number of protons in the observable universe. It is approximately 10^{80}. The initial mass of the spaceship (including propellant) would have to be much, much bigger than the mass of the observable universe.

So, that is not going to work. Chemical rockets cannot get us to 1% light speed.

Other engine types can produce higher exhaust velocities. A nuclear-thermal engine can produce an exhaust velocity of roughly 10,000 m/s, which is still much too low. With current technology, a typical ion drive has an exhaust velocity of about 40,000 m/s. It can't produce a high level of thrust, but it can produce a low level of thrust for a long time, using very little propellant.

With $V_E = 4 \times 10^4$ m/s, the rocket equation gives us roughly $M_0 = 4 \times 10^{37}$ kg. Now we're down to about 20 million times the mass of the Sun.

An ion drive won't get us to 1% of light speed.

The rocket equation implies that the exhaust velocity should be relatively close to the desired ΔV. To get to 1% of light speed, we need an engine with an exhaust velocity that is close to 1% of light speed.

Let's consider a few other proposed technologies for interstellar travel: light sails, nuclear-fusion engines and antimatter engines.

A light sail uses reflected light as a form of propulsion. Instead of throwing matter out the back at a high velocity, it simply reflects light. It requires an external source of light, such as a giant laser, and the light source must maintain a high intensity beam for a long time. The sail would have to be very big and very strong for its size, which makes it a difficult engineering problem. The energy transfer would become less efficient with increasing distance. Finally, a light sail would only work in one direction, and it wouldn't have brakes. It could only make a one-way trip, and it wouldn't be able to stop at its destination.

Project Daedalus proposed a rocket powered by small nuclear fusion explosions. (See source 56.) The spacecraft would have an initial mass of 54,000 metric tons, including 50,000 tons of nuclear fuel. It would carry a payload of 500 tons — less than 1% of the initial mass. It would reach 12% of light speed after four years, and then cruise toward its destination for 46 years, arriving at Barnard's star (a nearby red dwarf) after 50 years. It would not return to the Earth. It would deploy probes, and then send back information by a powerful focused radio signal.

At $1000/kg, it would cost $54 billion just to put the materials for the spacecraft into LEO. Of course, the actual construction costs would be much higher. What would humanity get for that investment? At best, some data about a nearby star. Project Daedalus was not designed to carry human passengers. 500 tons seems like a lot of mass, but it probably wouldn't be enough to support human existence for 50 years. Of course, there's not much point sustaining human life on a one-way suicide mission.

The antimatter drive is another proposed method of propulsion for reaching very high speeds. Essentially, antimatter and matter are combined in the engine, converting matter into energy. There are many problems with this idea, including how to create antimatter in large quantities, how to safely store it, and how to harness the energy of the reaction.

In general, a high exhaust velocity requires a very high energy level, at which dangerous and destructive radiation is created, such as X-rays, gamma rays and high-energy subatomic particles. The spacecraft and its crew (if manned) must be protected from this radiation. The higher the energy level, the more shielding is required.

Besides propulsion, there are other problems with high-speed travel.

One is the problem of how to slow down when we reach our destination (or preferably, before we reach it). With any known type of propulsion, this would require more propellant, and thus more mass. It could also be dangerous to travel into our exhaust gases.

There is also the danger of hitting something along the way. At a high speed, even dust particles are dangerous. The mass of a dust particle is about 1×10^{-9} kg. At 1% light speed, it would have 4,500 joules of kinetic energy, concentrated in a very small area. A typical rifle bullet has about 2,000 joules of kinetic energy. So, if you hit a dust particle at 1% light speed, it would do much more damage than a rifle bullet.

There is also the problem of space radiation, which was described in Asteroid Mining. On the Earth, we are protected from radiation by the magnetosphere and the atmosphere. In space, we would need shielding to protect us.

Ideally, for safe interstellar travel, we'd want to be inside a megastructure with a thick hull, or even inside an asteroid, under meters of rock. That would protect us from both collisions and radiation. Of course, such a vehicle would have an enormous mass, and it would require a huge amount of energy to accelerate to a high speed.

There is also the problem of maintaining life in space for hundreds or even thousands of years.

It is not possible to replicate the Earth's biosphere in a small volume. The poorly named Biosphere 2 was an attempt to create a self-sustaining closed ecosystem in a roughly 3-acre enclosure. It failed. A self-sustaining ecosystem would probably require something like the O'Neill cylinder described in Megastructures. The mass of our hypothetical cylinder was 3 trillion kg. That is 30 million times more massive than the Space Shuttle orbiter. Perhaps we could get away with a mere 1 trillion kg, or even 500 billion kg, but it would still be very massive.

Without a self-sustaining ecosystem, life would depend on technology and supplies. The mission would have to find a habitable planet before the supplies run out or the technology fails.

Let's not forget about gravity, or pseudo-gravity. Pseudo-gravity would be required to maintain a healthy human population and biosphere. That requirement would add yet more mass, energy and complexity to our hypothetical interstellar ark.

During the voyage, we'd need to maintain advanced technology for hundreds or thousands of years without access to an industrial economy. Imagine getting halfway to Proxima Centauri, and then having the life support system fail because of a faulty rubber gasket. We'd need to be able to fix, make or replace every component of the spaceship.

Modern technology is very complex. For that reason, it is somewhat fragile. A complex machine depends on a large number of parts. Even if each part is very reliable, as you increase the number of parts, it becomes almost certain that some part will fail.

When complex technology fails, it is hard to fix. These days, almost no one tries to fix their own television set, car or even washing machine. You need special parts, tools and knowledge. Gone are the days when a handyman could fix most of the technology in a typical home. Without access to an industrial economy, advanced technology is hard to maintain.

The typical lifespan of a modern machine is about 10 to 15 years. Can we make technology that lasts for hundreds or thousands of years? Perhaps, but not with conventional materials and methods. There is no known way of doing it. If we tried, we wouldn't know whether we had succeeded for a long time.

We'd also have to maintain social cohesion on a crowded spaceship for multiple generations. No major society has existed on the Earth for more than 400 years without at least one major social upheaval, such as a revolution.

Suppose that we solve all of those problems, and we discover a habitable planet that is not occupied by another technological civilization. Imagine a planet like the Earth 10 million years ago. It has plants and animals, oceans and polar ice caps. We can breathe the air. Maybe we can even eat the wildlife. It is the ideal planet for human colonization.

Now what?

We can't just transplant industrial civilization to this new planet. Even if we had a significant amount of starter capital, we would need to return to a simpler way of life, such as subsistence agriculture or hunting and gathering, while we slowly build up the population and economy to levels that can support industrial civilization. We'd have to hope that Terran organisms (including ourselves) can successfully compete with the native flora and fauna, despite not having evolved on the planet. We would begin a desperate struggle to exist as invasive species in a new environment.

Even if we survived and prospered, we might never be able to recreate industrial civilization. We might get stuck in a cycle of civilizational expansion and collapse. Ancient agricultural civilizations often collapsed, due to population growth and environmental destruction. In Central and South America, agricultural civilizations arose and collapsed over and over again, never advancing beyond neolithic agriculture, until

Europeans arrived. What if the evolutionary pathway to industrial civilization is hard to find, or depends on special conditions? We might never go down it again. There is no guarantee that human beings will always create something like modern civilization.

Given all the challenges of interstellar travel, and the dubious rewards, it seems very unlikely that any civilization would attempt it, unless they had magical technology, such as energy-cheap, faster-than-light space travel. Science fiction can imagine such technology, but there is no reason to believe that it will ever exist.

Social Filters

Interstellar colonization would be a global megaproject. It would require a huge investment. That investment would come from the people and resources of the Earth. Humanity would need to cooperate, at a global scale, toward a goal that would not come to fruition for thousands of years. That is far beyond anything our species is currently capable of.

To the people of the Earth, interstellar colonization would have enormous costs and no benefits. If it succeeded, it would only benefit the colonists. Their descendants would inherit a new world and create a new civilization. They might even evolve into a new species. Those left behind on the Earth (the other 99.999999% of humanity) would not benefit biologically from the venture, but they would have to contribute to it.

We reproduce as individuals. For that reason, biological purpose resides at the individual level, not at the level of society, species or biosphere. We can create collective values, but collective values are derived from individual values. They represent opportunities to benefit by cooperating. They don't reflect a shared purpose.

Interstellar colonization would require a huge collective effort. Thus, it would require the collective will to motivate that effort. Humanity would have to collectively *want* to colonize the stars. Why would they? Why would people agree to contribute resources and labor to a project that would not benefit them or their descendants?

It is unlikely that humanity will ever develop the collective will to colonize the stars. The same is true of any similar alien species.

Even if we had the global cooperation necessary for such a megaproject, there are many problems on the Earth that we could (and should) solve before we worry about interstellar colonization. Conversely, if we can't develop the global cooperation necessary to solve much easier problems, such as regulating our population, or switching from fossil fuels to renewable energy, then we will never go to the stars.

The Star Trek Worldview

When I was young, I loved watching Star Trek. I mean the original series, not the various sequels and spin-offs. I have the DVD collection, and I still watch episodes occasionally. For a TV show, it is both entertaining and thought-provoking, which is a rare combination. In this chapter, I will explore the strange and often contradictory worldview of Star Trek.

Star Trek is a mish-mash of different entertainment genres that were popular when it was made. It is a "space Western" about explorers and pioneers on a "final frontier". It is a military drama-comedy about the (male) camaraderie of military life. It is a boy's adventure story, with strange places, creatures and predicaments. Many episodes involve a mystery, and some have the format of a classic detective show. It has verbal humor and some situational comedy. A love story is often part of the plot. The creators of Star Trek threw almost every TV trope into the show. Somehow, it all stuck together and worked.

Star Trek was intended to convey certain ideas about humanity and progress. It portrays a future of racial harmony, sexual equality, scientific progress, powerful technology, prosperity and adventure. Star Trek has an essentially progressive view of history: that humanity will advance not only technologically but also morally. It imagines a future in which humanity has moved past hatred and violence. This brighter future is not inevitable, however. We must struggle toward it, and make the right choices to attain it. But it is attainable.

We're human beings with the blood of a million savage years on our hands, but we can stop it! We can admit that we're killers, but we're not going to kill today.

That's all it takes... knowing that we're not going to kill today.

— Kirk in <u>A Taste of Armageddon</u>

Star Trek also reflects the issues and norms of its time. It was created in the mid-1960s, a time of rapid material and intellectual progress. The preceding 50 years had seen the greatest expansion of science, technology and prosperity in human history. The United States was leading the world in every respect: science, technology, prosperity, military power and popular culture. Americans were naturally optimistic and confident about the future and their own cultural superiority.

However, there were some dark clouds on the horizon. Americans were fighting in Vietnam. There was political conflict over the war and various social issues. The threat of nuclear war was always in the background. There was growing concern about the problems of modern civilization, such as pollution, environmental destruction, population growth, and the depletion of natural resources.

There were also psychological and philosophical problems. Modern civilization generated prosperity, but it also caused alienation and anxiety. Would we become just cogs in the industrial machine, controlled by technocrats and computers? Modern science undermined traditional beliefs about human nature and our place in the cosmos. This raised questions about purpose and meaning. The West was also going through the sexual revolution, due to changed economic conditions and the birth control pill. A new way of life was emerging, which involved low fertility, extended adolescence, women in the workforce, and sexual freedom. It was a time of sexual and social change and experimentation.

Star Trek reflects the culture of its time and place. As a science fiction show, it specifically deals with hopes and fears about the future. Overall, it has a liberal/progressive message, but it isn't preachy. It doesn't present a morally black and white world. Instead, its characters often struggle with moral ambiguities and conflicts. In that way, Star Trek is quite intellectual, especially for an entertaining TV show.

Despite being one of the smartest TV shows of all time, Star Trek contains many contradictions, which were mostly overlooked by its creators and fans. In the rest of this chapter, I will describe the major contradictions of the Star Trek worldview.

Race and Group Conflict

In the Star Trek future, humanity has risen above its violent past of racial prejudice and group conflict. People of different races and ethnicities live and work together in peace and harmony, as part of a united humanity.

However, Star Trek also portrays a galaxy full of racialized group conflict. The Federation is expanding its (supposedly benevolent and peaceful) rule. By doing so, it has come into conflict with the Klingons and the Romulans, who are portrayed as *evil* alien races. They make emotionally compelling enemies because they are humanoid, not true aliens. (The silicon rock monster of The Devil in the Dark is a true alien.) The Klingons and Romulans are foreign, but still human in appearance and behavior.

Star Trek uses racialized group conflict as a plot device and a way to make the show interesting. Essentially, Star Trek is about a military vessel on a mission of exploration and conquest. That mission naturally brings it into conflict with alien races.

Stories need conflict to be interesting. If every race in the galaxy was "enlightened" and peaceful, what would the Enterprise do? Go around the galaxy hugging everyone? That would be boring. We evolved to fight, both as individuals and as groups, so we like stories about conflict.

Star Trek caters to the audience's thirst for violence in many other ways. Most episodes have some type of violence, and many have multiple types, such as hand-to-hand combat, shoot-outs and space battles. Kirk tries to avoid violence, and yet it happens in almost every show.

Star Trek does not show us a utopian future in which we have transcended violence and racial divisions. Instead, it portrays racialized group conflict at a larger scale, and it rationalizes violence as serving a higher moral purpose.

Ethnic Identity and Unity

As it does with race, Star Trek has a split-brain vision of ethnic identity.

The crew of the Enterprise includes people of different ethnic backgrounds, such as Scotty (Scottish), Sulu (Japanese), Chekov (Russian), Kirk (American) and Uhura (East African). But how and why do these ethnic identities exist in the Star Trek future?

Supposedly, humanity has transcended the national divisions of the past. The Earth has a united society with a single government. Presumably, there is global freedom of movement. Under those conditions, the world would become a biological and cultural melting pot. People and ideas would mix. Current racial categories and ethnic identities would not persist.

And yet, ethnic identities have somehow persisted in the Star Trek future. Scotty has a Scottish accent, and he likes to drink Scotch whiskey. Chekov has a Russian accent, and he is very proud of the accomplishments of the Russian people. Uhura speaks Swahili. Why would these ethnic identities and differences exist in a world without national boundaries?

The creators of Star Trek probably didn't give it much thought. The show reflected the United States of its time, which was populated mostly by the descendants of European immigrants. Many people had a secondary ethnic identity, such as Italian, Polish or Irish. Those secondary ethnic identities were not very important or divisive. The situation in Star Trek was similar. The ethnic identities didn't divide people. The ethnic differences were trivial aesthetic preferences, such as preferring whiskey to vodka.

Although liberals get a little *too* excited about ethnic restaurants, ethnic diversity can make a society more interesting. However, the "getting to know each other" phase doesn't last forever. When it's over, we've either segregated into somewhat hostile subcultures, or we've melted together into a single culture. Ethnic identity either ceases to exist, or it ceases to be benign. Ethnic diversity might have been the spice of life in late 20th century America, but not in Northern Ireland, Cyprus, Jerusalem, etc. In the long run, ethnic identities only persist if people are divided.

If humanity ever does form a single, global society without national boundaries, a global monoculture will emerge. However, there will also be subcultures within that culture. In the absence of ethnic identity, people create other group identities. In the 1980s, American teenagers organized themselves into subcultures based on musical tastes, clothing, interests and attitudes. There were punks, preppies, metal-heads, nerds, jocks, etc. Today, people self-organize on the internet around ideologies and shared interests. People naturally seek a balance between conformity and individuality, which typically involves creating somewhat exclusive group identities. Instead of ethnic identities, we can imagine a Star Trek crew with members of different ideological, religious or aesthetic subcultures.

Star Trek's confused vision of ethnic identity is an attempt to have it both ways: have ethnic diversity to make things interesting, but without the segregation that creates and maintains ethnicity.

Sex Roles

Star Trek envisions a society that is sexually egalitarian in some ways, but still has traditional/biological sex roles.

The main characters are male: Kirk, Spock, McCoy and Scotty. They have traditionally masculine jobs: captain, scientist, doctor and engineer. Most female characters have more feminine jobs, such as communications officer or nurse. Among the secondary characters, there are some female doctors and scientists, but they play their roles in a feminine way.

There is pronounced sexual dimorphism in appearance and behavior. Starfleet has different uniforms for men and women: shirts and pants for men, short dresses and tights for women. Alien women are often dressed even more provocatively. Men have short hair, while women have long hair, which is often styled. In behavior, the men are masculine, and the women are feminine. Men are strong, stoic and brave. They are either heroes or power-hungry villains. Women are damsels in distress or seductive vixens.

In the Star Trek future, humanity has not transcended beauty standards. There are clear norms of male and female attractiveness. Most of the women are young and beautiful. Most of the men are physically fit and handsome. Chekov is more cute than handsome, but he was added to the cast to appeal to younger female viewers. (This was during the time of Beatlemania and The Monkees.)

Although Star Trek portrays a society that is sexually egalitarian in certain ways, it also has a very clear male | female distinction. The men are real men, the women are real women, and the Gorn are real Gorn.

Technology and Human Agency

In Star Trek, magical technology does all the boring jobs, but it does not take away the interesting ones. The Enterprise has a large crew. It is so large that the captain does not know every crew member personally. Despite the magical technology, humans are still necessary for the functioning of the Enterprise.

The magical technology breaks down occasionally, and it has to be fixed with the equivalent of wrenches and soldering irons. When enemy fire hits the ship, instrument panels short out, and must be repaired by hand. Modern technology is not like that. Men can't tinker with their cars anymore. Most of our technology requires specialized parts and equipment to repair. But in Star Trek, the magical technology of the future is like the cars and household appliances of the 1960s. A handyman can fix it himself with basic knowledge and simple tools.

This is not plausible, of course. It is not likely that the magical technology could be fixed by hand with a wrench and a soldering iron. But it is important to the show. Men are not just passengers, riding along with powerful technology, like fleas on a dog. They control it, and they are necessary for it to function. One recurring theme in Star Trek is that humans should rule over technology, not vice versa.

But what if technology had not just eliminated the boring jobs, such as taking out the garbage and making food? What if it could do *every* job better than a human? Why bring humans along at all? Why not just send robots to explore the final frontier?

As it does with race, sex and violence, Star Trek tries to have it both ways. It imagines magical technology that expands human agency, but does not make humans obsolete.

Star Trek explores dystopias in which computers control humans (or humanoid aliens) for their own good. It never explains how such dystopias can be avoided, or even why they are wrong per se. If technological progress continues unabated into the future, why wouldn't computers become better at making decisions than humans? Why would we reserve certain tasks for humans instead of machines?

Those are interesting questions to explore, but exploring them would not make good entertainment. We want to see people doing things. We want to see them solving problems that fully engage and test human abilities.

The jobs on the Enterprise are more interesting and engage more human abilities than the typical job today. Modern civilization has led to an extreme specialization of labor. Some people just run one machine, or create spreadsheets, or drive a bus down the same route every day, or make coffee, etc. Modern jobs are boring, because they are so specialized. They don't engage all of our mental and physical abilities.

When you come home from your boring job, you don't want to watch people doing boring jobs on TV. You want to see something interesting, such as handsome men fighting monsters and seducing beautiful women.

Eugenics | Dysgenics

In the Star Trek worldview, eugenics is morally wrong, and it has been rejected by humanity. In the Star Trek timeline, humanity fought the "eugenics wars" in the late 20th century. A race created by eugenics tried to take over the world, failed, and was exiled to space. That is the backstory of the episode Space Seed and the movie The Wrath of Khan. Eugenics is associated with evil.

However, everyone on the Enterprise is healthy, good-looking and intelligent. Is the crew a representative sample of the human population? Probably not. Presumably, they were selected for traits such as intelligence, health and cooperativeness. If it is acceptable and beneficial to select a ship's crew, why not select the human population in general?

Eugenics is necessary to maintain an advanced civilization for a long period of time. Mutations are always being added to the genome. To maintain a healthy population, mutations must be removed by selection of some kind (natural or artificial). In the absence of high childhood mortality (the ancestral condition), eugenics is necessary to maintain a healthy genome. Civilization also requires certain traits in the population, such as intelligence and responsibility. To be sustainable, a civilization must select for the traits that make it possible.

As usual, Star Trek tries to have it both ways. It portrays an advanced civilization in the distant future, populated by intelligent, healthy and beautiful people — without eugenics.

Futurism and Primitivism

Like most science fiction, Star Trek is as *primitivistic* as it is futuristic. It uses futuristic ideas, such as space travel, to put its characters into primitive situations.

In almost every episode, a few crew members (usually including Kirk) beam down to a strange planet, which luckily happens to be "Class M", meaning that it is almost identical to southern California in climate and vegetation. The atmosphere is breathable, and the alien inhabitants speak English, even if they have oddly colored skin and wear strange clothing. We can forgive a low-budget TV show for that.

After the landing party has beamed down, their powerful phaser weapons are taken away or rendered useless, forcing them to engage in hand-to-hand combat. Kirk is often seen rolling on the ground, shirt ripped, with a few minor scratches. Being more cerebral and less violent, Spock typically uses his "nerve pinch" to incapacitate enemies, but he can also use his Vulcan strength to throw them around. Kirk and Spock solve problems with their wits and physical abilities. When they use technology, they often need to fix it or make it themselves. Kirk also relies on his charisma and persuasion skills, including his ability to charm women.

Many episodes involve recapitulating the Earth's past in some way. In The Galileo Seven, a shuttle is stranded on a planet populated by giant cave-men, who attack the crew with stones and spears. In Arena, Kirk is teleported to a planet where he must fight an enemy ship's captain, using only weapons that

95

they create themselves. Kirk wins by creating a primitive gun. In <u>The Return of the Archons</u>, they visit a planet that has a society like the United States in 1850. In <u>This Side of Paradise</u>, they visit a colony whose inhabitants have a simple agricultural lifestyle. In <u>The Paradise Syndrome</u>, Kirk spends time with a tribe of Native Americans. There are many other examples. Even when the aliens are advanced, the setting is often archaic, as in <u>The Squire of Gothos</u>.

Science fiction does this because it works. Our brains are not adapted to modern civilization. Our ancestors had to move around in natural landscapes, solve physical problems, avoid dangers, hunt for food, and fight enemies at close range. Few of us would trade the comfort and security of modern civilization for that ancestral condition, but we still yearn for it in a way. When modern man is sitting comfortably on his couch, after a day of working at his safe, boring job, he wants to vicariously participate in those ancestral behaviors. So, science fiction often presents us with situations and problems that resemble the exciting parts of our ancestor's lives. That's why Jedis fight with swords in Star Wars. It makes no sense, but it is entertaining.

Utopian Dystopias

Star Trek presents us with a positive vision of humanity's future, but it isn't exactly utopian. If it was utopian, it would be boring. There would be nothing to do. Utopia, almost by definition, is the end of history. What people find appealing is not utopia itself, but the struggle toward utopia. In Star Trek, there is a positive direction for us to go in, but we're not anywhere near a final destination. Humanity is expanding, exploring, growing and progressing.

Star Trek explores many dystopias that arise from attempts to create utopias. So, although Star Trek has an essentially liberal/progressive message, it also has a deeper conservative/reactionary message: that the attempt to create a utopia will often result in a dystopia.

The Apple presents a typical Star Trek utopian dystopia. Kirk, Spock, McCoy and others beam down to explore a planet. It seems like a paradise at first, but there are hidden dangers. Three "redshirts" (disposable characters) are killed in various ways. Eventually, the landing party encounters the humanoid inhabitants of the planet: oddly colored people who speak English. The natives have a peaceful, primitive existence. They worship a deity named "Vaal". Vaal gives them eternal youth, free from hunger, disease and war. However, Vaal does not permit them to have sex or progress culturally. The natives live in a stagnant Eden.

To McCoy, this "paradise" is a monstrosity. Spock warns that intervening would violate the prime directive, which requires them to not interfere in the development of alien civilizations. (Of course, they routinely violate this directive.) As in many episodes, the Enterprise is being pulled toward the planet by a powerful tractor beam. To save his ship, Kirk must destroy the deity, which (of course) is actually a computer. By doing so, he releases the natives from their captivity in paradise.

This is not the typical Hollywood "good guys versus bad guys" narrative. It is more complex. The "bad guy" is a computer that was designed to create a benevolent, if authoritarian, utopia. Like many Star Trek episodes, The Apple raises philosophical questions. What is a utopia? Does progress have a destination? If so, what is it? Should we sacrifice freedom for peace? Should we trade love and sex for immortality? Should we liberate people from authoritarian rule? Or do they choose that fate by accepting it?

The episode doesn't really answer those questions, at least not explicitly. But it does raise them.

Many other episodes involve some type of dystopian utopia, in which there is a dilemma or trade-off between competing values. I will give a few examples.

In Mudd's Women, plain women are made attractive by a drug. This raises the question of whether reality matters, or only our experiences of it. Is a beautiful illusion better than an ugly reality?

In What Are Little Girls Made Of?, a rogue scientist discovers alien robot technology, which he uses to make himself immortal and to create beautiful female servants. The Enterprise arrives to search for him, and to rescue him if he is alive. His girlfriend (Nurse Chapel) is on the Enterprise. He wants to be with her, but he can't return to the human world, because he has become an android. The episode raises questions about what we now call "trans-humanism". Can the mind exist without the body, or in a new type of body? Should we try to transcend aspects of our humanity, such as aging and mortality? The episode is also about power, freedom and society. Is it better to reign in isolation, or be part of a community that limits your freedom?

In The Return of the Archons, a landing party (including the standard trio of Kirk, Spock and McCoy) beams down to a planet where a starship was lost in the past. They are supposed to investigate the missing starship, but not interfere in the planet's cultural development. However, they discover that the planet is run by a computer that demands the total subordination of individual will and self-interest to the collective. This raises questions about individualism versus collectivism, and utilitarianism versus individual rights. As usual, they break the prime directive and destroy the computer, thereby liberating the planet and escaping from it.

In these episodes and many others, Star Trek explores the dark side of progress and utopias.

There is one utopia that is mysteriously unexplored in Star Trek: the Earth. Star Trek takes place on humanity's expanding frontier, not on the Earth. What is the Earth like? If every problem on the Earth has been solved, what do its inhabitants do? Wouldn't the Earth be boring and stagnant, and thus another dystopian utopia?

Conclusion

The contradictions of Star Trek reveal problems with the humanist worldview. Humanism pretends to be scientific and rational, but it has an essentially religious view of humanity and nature.

Many of the contradictions arise out of the conflict between the humanist vision of Star Trek and the need to make it entertaining. A peaceful, prosperous utopia would be boring. It's hard to even imagine. We evolved to struggle and fight to survive. So, we find tales of struggle and conflict interesting. Love and sex are interesting too, but there must be a problem to solve. The human brain is a problem-solving machine, so we find it interesting to watch other people solve problems. We also evolved to act in the world: to use our physical abilities as well as our mental abilities. So, we like to watch people acting physically. To be entertaining, Star Trek must appeal to human nature, but it also presents a vision in which we transcend certain aspects of human nature.

The conflict between emotion and logic is a recurring theme in Star Trek, with McCoy and Spock representing the two sides. Emotion is not really in opposition to logic, and logic isn't what Spock represents anyway. Spock represents *rationality*, which is broader than logic. Essentially, rationality is the use of thought, rather than intuition, to solve problems. We can't be fully rational, because thought depends on intuition. However, the character of Spock represents something important.

As we progress, we move further away from the ancestral condition, to which our instincts are adapted. So, we must rely more and more on thought and explicit knowledge, rather than our instincts. We need to become increasingly self-critical and self-controlled. Spock is not devoid of emotion. He has his emotions under conscious control, which is a type of emotional state. To adapt to modern civilization, we need to become more like Spock. We need to be more rational, and find ways to control our emotions.

Star Trek struggles with this conflict, but like the other conflicts in Star Trek, it is never resolved. The Vulcans are portrayed as superior in some ways, but inferior in others: another dystopian utopia that we must avoid. We must retain our humanity, while suppressing or transcending certain aspects of it.

Humanity is the sacred value in Star Trek, but what is this sacred value? Does it include our capacity for hatred and violence? Our greed and selfishness? Or are those not essential aspects of humanity, but mere blemishes on the human spirit? Star Trek does not provide answers, but it does raise the questions.

That's why I like Star Trek, despite its contradictions. It raises interesting questions.

Progress is Not Inevitable

These days, most people view history as a "march of progress". They believe that progress is the natural direction of history. There might be setbacks and obstacles along the way, but progress is inevitable in the long run.

This view of history is an illusion created by two errors of perspective. One is that we naturally view ourselves as good, so we see our history as progress, because it produced us. The other is that we are one of history's winners. If we weren't, we wouldn't be here. When we look back on our history, we see "progress" toward ourselves. Thus, it seems that history is on our side, and we are destined to succeed. But history has far more losers than winners. Looking backward, we are guaranteed to be winners by the anthropic principle. Looking forward, we are very likely to be losers.

Life consists of an enormous number of experiments, most of which fail. That is true for individuals, species and civilizations. We should not assume that the future belongs to us. The most likely outcome for our civilization is collapse. The most likely outcome for our species is extinction without descendants.

There is nothing wrong with optimistic fantasies per se. They become a problem when they are mistaken for realistic predictions or pragmatic prescriptions. Optimism can blind us to real problems and solutions. Even pessimistic fantasies can blind us. We can waste time worrying about nonexistent threats, while real dangers are ignored.

The modern attitude toward technology is based on ignorance. Most people have little knowledge of physics, economics or computing science. In the modern worldview, technology has a quasi-religious significance. It is a blank screen onto which people project their fantasies and fears.

Some people believe that technology will create a paradise on Earth. There is even a technological version of heaven: having your mind uploaded into a computer or a robot body to avoid death. Some people project ancestral fears of monsters and enemies onto technology. They worry about robots taking over the world.

In reality, technology is highly constrained by physics and economics. We can do a lot within those limits, but technology is not magic.

The real dangers of modern technology come from the expansion of human agency, not from malevolent robots. Modern technology gives us powers that we didn't evolve to wield, and choices that we didn't evolve to make

We live in a very dangerous time, and a time of great change. Our civilization is not on a natural trajectory of progress. It is on a natural trajectory of explosive growth followed by catastrophic collapse. It is like a forest fire that grows by consuming its fuel, but will eventually burn itself out. There has been enormous progress in recent history, but it has not produced a stable civilization.

Realism and pragmatism, not blind optimism, give us the best chance of avoiding that outcome, and having a future.

Sources

1. Square/Cube Law
   https://en.wikipedia.org/wiki/Square%E2%80%93cube_la
   w

2. <u>Ringworld</u> by Larry Niven
   https://en.wikipedia.org/wiki/Ringworld

3. Dyson Sphere
   https://en.wikipedia.org/wiki/Dyson_sphere

4. Space Elevator
   https://en.wikipedia.org/wiki/Space_elevator

5. O'Neill Cylinder
   https://en.wikipedia.org/wiki/O%27Neill_cylinder

6. <u>Space Launch to Low Earth Orbit: How Much Does It
   Cost?</u> by Aerospace Security
   https://aerospace.csis.org/data/space-launch-to-low-earth-
   orbit-how-much-does-it-cost/

7. International Space Station
   https://en.wikipedia.org/wiki/International_Space_Station#
   Cost

8. Cost of ISS
   https://oig.nasa.gov/wp-content/uploads/2024/02/IG-22-
   005.pdf

9. Tsiolkovsky rocket equation
   https://en.wikipedia.org/wiki/Tsiolkovsky_rocket_equation

10. ΔV
    https://en.wikipedia.org/wiki/Delta-v

11. Saturn V Rocket
https://en.wikipedia.org/wiki/Saturn_V

12. Space Shuttle
https://en.wikipedia.org/wiki/Space_Shuttle

13. Space Shuttle External Tank
https://en.wikipedia.org/wiki/Space_Shuttle_external_tank

14. Space Shuttle Challenger Disaster
https://en.wikipedia.org/wiki/Space_Shuttle_Challenger_di
saster

15. Teacher in Space Project
https://en.wikipedia.org/wiki/Teacher_in_Space_Project

16. Orbital Ring
https://en.wikipedia.org/wiki/Orbital_ring

17. Linear Motor
https://en.wikipedia.org/wiki/Linear_motor

18. Millennium Falcon
https://en.wikipedia.org/wiki/Millennium_Falcon

19. Space Capsule
https://en.wikipedia.org/wiki/Space_capsule

20. Space Plane
https://en.wikipedia.org/wiki/Spaceplane

21. Space Shuttle Columbia Disaster
https://en.wikipedia.org/wiki/Space_Shuttle_Columbia_dis
aster

22. SpaceX Starship
https://en.wikipedia.org/wiki/SpaceX_Starship

23. Astronaut Mass Balance for Long Duration Missions by
Michael K. Ewert and Chel Stromgren

https://ntrs.nasa.gov/api/citations/20190027563/downloads/20190027563.pdf

24. Radioisotope Thermoelectric Generator
https://en.wikipedia.org/wiki/Radioisotope_thermoelectric_generator

25. Radiation Sources and Doses
https://www.epa.gov/radiation/radiation-sources-and-doses

26. Radiation Exposure Comparisons with Mars Trip Calculation
https://science.nasa.gov/resource/radiation-exposure-comparisons-with-mars-trip-calculation/

27. Health Effects of Spaceflight
https://en.wikipedia.org/wiki/Effect_of_spaceflight_on_the_human_body

28. Effective Temperature of the Earth
https://en.wikipedia.org/wiki/Stefan%E2%80%93Boltzmann_law#Effective_temperature_of_the_Earth

29. 16 Psyche
https://en.wikipedia.org/wiki/16_Psyche

30. Rocket Man
https://en.wikipedia.org/wiki/Rocket_Man_(song)

31. Uncanny Valley Effect
https://en.wikipedia.org/wiki/Uncanny_valley

32. Ethical Issues in Advanced Artificial Intelligence by Nick Bostrom
https://nickbostrom.com/ethics/ai

33. Paperclip Maximizer
https://www.lesswrong.com/tag/squiggle-maximizer-formerly-paperclip-maximizer

34. The Restaurant at the End of the Universe by Douglas
    Adams
    https://en.wikipedia.org/wiki/The_Restaurant_at_the_End_
    of_the_Universe

35. Cognitive Wheels: The Frame Problem of AI by Daniel
    Dennett
    https://www.researchgate.net/publication/225070451_Cogn
    itive_Wheels_The_Frame_Problem_of_AI

36. The Frame Problem in the Stanford Encyclopedia of
    Philosophy
    https://plato.stanford.edu/entries/frame-problem/

37. Technological Singularity
    https://en.wikipedia.org/wiki/Technological_singularity

38. The Coming Technological Singularity by Vernor Vinge
    https://ntrs.nasa.gov/citations/19940022856

39. Fanged Noumena by Nick Land
    https://en.wikipedia.org/wiki/Fanged_Noumena

40. The Fermi Paradox
    https://en.wikipedia.org/wiki/Fermi_paradox

41. Milky Way Galaxy
    https://en.wikipedia.org/wiki/Milky_Way

42. G-Type Main-Sequence Star
    https://en.wikipedia.org/wiki/G-type_main-sequence_star

43. K-Type Main-Sequence Star
    https://en.wikipedia.org/wiki/K-type_main-sequence_star

44. Habitable Zone
    https://en.wikipedia.org/wiki/Habitable_zone

45. Snowball Earth
https://en.wikipedia.org/wiki/Snowball_Earth

46. Permian-Triassic Extinction
https://en.wikipedia.org/wiki/Permian%E2%80%93Triassic
_extinction_event

47. Cretaceous-Paleogene Extinction
https://en.wikipedia.org/wiki/Cretaceous%E2%80%93Pale
ogene_extinction_event

48. Abiogenesis
https://en.wikipedia.org/wiki/Abiogenesis

49. Prokaryote
https://en.wikipedia.org/wiki/Prokaryote

50. Eukaryote
https://en.wikipedia.org/wiki/Eukaryote

51. Proxima Centauri
https://en.wikipedia.org/wiki/Proxima_Centauri

52. Voyager 1
https://en.wikipedia.org/wiki/Voyager_1

53. Eddington Number
https://en.wikipedia.org/wiki/Eddington_number

54. Ion Thruster
https://en.wikipedia.org/wiki/Ion_thruster

55. Light Sail
https://en.wikipedia.org/wiki/Solar_sail

56. Project Daedalus
https://en.wikipedia.org/wiki/Project_Daedalus

57. Antimatter Drive
https://en.wikipedia.org/wiki/Antimatter_rocket

58. Biosphere 2
https://en.wikipedia.org/wiki/Biosphere_2

59. Star Trek
https://en.wikipedia.org/wiki/Star_Trek:_The_Original_Ser
ies

60. A Taste of Armageddon
https://en.wikipedia.org/wiki/A_Taste_of_Armageddon

61. The Devil in the Dark
https://en.wikipedia.org/wiki/The_Devil_in_the_Dark

62. Space Seed
https://en.wikipedia.org/wiki/Space_Seed

63. The Wrath of Khan
https://en.wikipedia.org/wiki/Star_Trek_II:_The_Wrath_of
_Khan

64. The Galileo Seven
https://en.wikipedia.org/wiki/The_Galileo_Seven

65. Arena
https://en.wikipedia.org/wiki/Arena_(Star_Trek:_The_Orig
inal_Series)

66. The Return of the Archons
https://en.wikipedia.org/wiki/The_Return_of_the_Archons

67. This Side of Paradise
https://en.wikipedia.org/wiki/This_Side_of_Paradise_(Star
_Trek:_The_Original_Series)

68. The Paradise Syndrome
https://en.wikipedia.org/wiki/The_Paradise_Syndrome

69. The Squire of Gothos
https://en.wikipedia.org/wiki/The_Squire_of_Gothos

70. The Apple
https://en.wikipedia.org/wiki/The_Apple_(Star_Trek:_The_Original_Series)

71. Redshirt
https://en.wikipedia.org/wiki/Redshirt_(stock_character)

72. Mudd's Women
https://en.wikipedia.org/wiki/Mudd%27s_Women

73. What Are Little Girls Made Of?
https://en.wikipedia.org/wiki/What_Are_Little_Girls_Made_Of%3F